STUDENT'S STUDY G

Elementary Algebra

STUDENT'S STUDY GUIDE

Elementary Algebra

DUGOPOLSKI

Grace Hu
Southeastern Louisiana University

ADDISON-WESLEY PUBLISHING COMPANY
Reading, Massachusetts • Menlo Park, California • New York
Don Mills, Ontario • Wokingham, England • Amsterdam • Bonn
Sydney • Singapore • Tokyo • Madrid • San Juan

Reproduced by Addison-Wesley from camera-ready copy supplied by the author.

ISBN 0-201-53349-9

2 3 4 5 6 7 8 9 10 - BA - 95 94

TO THE STUDENT:

This student study guide is a supplement to the text
Elementary Algebra by Mark Dugopolski. This guide is designed to
assist you in developing a clear understanding of the subject
matter through practice.

Each topic begins with an overview of the subject, followed
by examples and exercises along with detailed solutions.
A comprehensive test concludes each chapter. Diligent practice is
important in learning algebra, and this study guide provides
hundreds of problems with their solutions.

It is my sincere hope that you will find this study guide
beneficial and enjoyable.

Grace Hu

CONTENTS

CHAPTER 1   • From Arithmetic to Algebra   1
           • Tests   17

CHAPTER 2   • Linear Equations and
             Inequalities in One Variable   21
           • Tests   34

CHAPTER 3   • Exponents and Polynomials   37
           • Tests   51

CHAPTER 4   • Factoring   57
           • Tests   80

CHAPTER 5   • Rational Expressions   86
           • Tests   115

CHAPTER 6   • Powers and Roots   123
           • Tests   150

CHAPTER 7   • Quadratic Equations   158
           • Tests   170

CHAPTER 8   • Linear Equations in Two Variables   177
           • Tests   202

CHAPTER 9   • Systems of Equations and
             Inequalities in Two Variables   209
           • Tests   238

CHAPTER 1

FROM ARITHMETIC TO ALGEBRA

1.1 FRACTIONS

OBJECTIVES:

*(A) To express a fraction as another equivalent fraction.
*(B) To express a fraction as a decimal or percentage, and vice versa.
*(C) To add, subtract, multiply, and divide fractions.

** REVIEW

*(A) Express a fraction as another equivalent fraction.
Always multiply the denominator and numerator by exactly
the same non-zero factor.

Examples:

$$a. \ \frac{15}{14} = \frac{?}{42} \quad\quad b. \ \frac{15}{14} = \frac{30}{?} \quad\quad c. \ \frac{15}{14} = \frac{?}{420} \quad\quad d. \ \frac{15}{14} = \frac{75}{?}$$

Solutions:

$$a. \ \frac{15}{14} = \frac{?}{42}$$

14 (Denominator) x factor = 42.
Therefore, the factor = 42 ÷ 14 = 3

$$\frac{15}{14} = \frac{15 \times 3}{14 \times 3} = \frac{45}{42}$$

$$b. \ \frac{15}{14} = \frac{30}{?}$$

15 (Numerator) x factor = 30.
Therefore, the factor = 30 ÷ 15 = 2

$$\frac{15}{14} = \frac{15 \times 2}{14 \times 2} = \frac{30}{28}$$

Practice with the remaining problems.

$$c. \ \frac{15}{14} = \frac{450}{420} \ , \quad\quad d. \ \frac{15}{14} = \frac{75}{70}$$

(Hint: The factors for c and d are 30 and 5, respectively)

*(B) Express a fraction as a decimal or percentage, and vice versa.

● Change the **fraction** to a **decimal**.

Examples: a. $\dfrac{7}{14}=0.5$ b. $\dfrac{15}{4}=3.75$

● Change the **fraction** to a **percentage**.
(After changing the **fraction** to a **decimal**, **multiply** the decimal by **100%**)

Example:
   7/14 = ?%
   Change 7/14 to a decimal ---> 7/14 = 0.5
   Multiply the decimal by 100% --> 0.5 x 100% =50%

● Change a decimal to a fraction.
(If the decimal has x decimal places, then place x zeros after 1 in the denominator.)

Examples:
   a. 1.23 = 123/100   (1.23 has 2 decimal places)
   b. 0.00123 = 123/100000   (0.00123 has 5 decimal places)
   c. 3.45 = 345/100   (3.45 has 2 decimal places)

● Change a percentage to a fraction. (x% means x/100)
● Change a percentage to a decimal.

Examples:
   a. 25% = 25/100 = 1/4 = 0.25
   b. 1.23% = 1.23/100 = 123/10000 = 0.0123
   c. 0.75% = 0.75/100 = 0.0075
   (**Hint**: Divide the number by 100 and simplify)

*(C) Add, subtract, multiply, and divide fractions.

● Add or subtract fractions
   a. Find the <u>L</u>east <u>C</u>ommon <u>D</u>enominator (LCD)
   b. Convert each fraction to an equivalent fraction having the least common denominator, then add or subtract the numerators.

Examples:

   a. $\dfrac{3}{4}+\dfrac{2}{5}=?$ *b.* $\dfrac{3}{8}-\dfrac{1}{9}=?$ *c.* $\dfrac{14}{53}+\dfrac{8}{53}=?$ *d.* $\dfrac{2}{3}-\dfrac{3}{5}=?$

Solutions:

   a. $\dfrac{3}{4}+\dfrac{2}{5}=\dfrac{3\times5}{4\times5}+\dfrac{2\times4}{5\times4}=\dfrac{15}{20}+\dfrac{8}{20}=\dfrac{23}{20}=1\dfrac{3}{20}$

$b.\ \dfrac{3}{8} - \dfrac{1}{9} = \dfrac{3\times9}{8\times9} - \dfrac{1\times8}{9\times8} = \dfrac{27-8}{72} = \dfrac{19}{72}$

$c.\ \dfrac{14}{53} + \dfrac{8}{53} = \dfrac{22}{53} \qquad d.\ \dfrac{2}{3} - \dfrac{3}{5} = \dfrac{1}{15}$

● Multiply two fractions.
 Multiply the numerators. The product is the new numerator.
 Multiply the denominators. The product is the new denominator.

Examples:

$a.\ \dfrac{1}{3} \times \dfrac{2}{5} = \dfrac{1\times2}{3\times5} = \dfrac{2}{15} \qquad b.\ \dfrac{1}{2} \times \dfrac{3}{7} \times \dfrac{4}{9} = \dfrac{1\times3\times4}{2\times7\times9} = \dfrac{12}{126} = \dfrac{6}{63}$

● Divide one fraction by another fraction.
 Invert the second fraction and multiply the two fractions.

Examples:

$a.\ \dfrac{3}{5} \div \dfrac{7}{4} = \dfrac{3}{5} \times \dfrac{4}{7} = \dfrac{3\times4}{5\times7} = \dfrac{12}{35} \qquad b.\ \dfrac{2}{15} \div \dfrac{3}{15} = \dfrac{2}{15} \times \dfrac{15}{3} = \dfrac{2\times15}{15\times3} = \dfrac{2}{3}$

◆◆◆◆◆

**<u>EXERCISES</u>

1. Fill in the missing values.

$$\dfrac{30}{50} = \dfrac{?}{5} = \dfrac{15}{?} = \dfrac{21}{?} = \dfrac{?}{150}$$

2. Perform the following operations.

$a.\ \dfrac{1}{3} + \dfrac{2}{5} = ? \qquad b.\ \dfrac{2}{3} - \dfrac{1}{8} = ? \qquad c.\ 2 + \dfrac{1}{2} + \dfrac{1}{7} = ? \qquad d.\ \dfrac{2}{5} \times \dfrac{3}{7} = ?$

$e.\ \dfrac{9}{10} \div 18 = ? \qquad f.\ 1 + \dfrac{1}{2} + \dfrac{1}{3} + \dfrac{1}{4} = ? \qquad g.\ 1 \div \dfrac{1}{2} \div \dfrac{1}{3} \div \dfrac{1}{4} = ?$

   Hint for : (f). LCD = 2·3·4 = 12
            (g). Perform division from left to right.

3. Fill in the missing values.

<u>fraction</u>	<u>decimal</u>	<u>percentage</u>
8/10	?	?
?	2.5	?
?	?	15%

   (Note : **a/b means a is divided by b**)
◆◆◆◆◆

3

** SOLUTIONS TO EXERCISES

1. Fill in the missing values.

$$\frac{30}{50} = \frac{3}{5} = \frac{15}{25} = \frac{6}{10} = \frac{21}{35} = \frac{90}{150} = \frac{180}{300} = \frac{12}{20}$$

2. Perform the following operations.

a. $\dfrac{1}{3} + \dfrac{2}{5} = \dfrac{1 \times 5 + 2 \times 3}{15} = \dfrac{5+6}{15} = \dfrac{11}{15}$ b. $\dfrac{2}{3} - \dfrac{1}{8} = \dfrac{16-3}{24} = \dfrac{13}{24}$

c. $2 + \dfrac{1}{2} + \dfrac{1}{7} = 2 + \dfrac{1 \times 7 + 1 \times 2}{14} = 2\dfrac{9}{14}$ d. $\dfrac{2}{5} \times \dfrac{3}{7} = \dfrac{2 \times 3}{5 \times 7} = \dfrac{6}{35}$

e. $\dfrac{9}{10} \div 18 = \dfrac{9}{10} \times \dfrac{1}{18} = \dfrac{9}{180} = \dfrac{1}{20}$ f. $1 + \dfrac{1}{2} + \dfrac{1}{3} + \dfrac{1}{4} = \dfrac{25}{12} = 2\dfrac{1}{12}$

g. $1 \div \dfrac{1}{2} \div \dfrac{1}{3} \div \dfrac{1}{4} = (1 \times 2) \div \dfrac{1}{3} \div \dfrac{1}{4} = (2 \times 3) \div \dfrac{1}{4} = 6 \times 4 = 24$

3. Fill in the missing values.

fraction	decimal	percentage
8/10	0.8	80%
25/10	2.5	250%
15/100=3/20	0.15	15%

♦♦♦♦♦

1.2 THE REAL NUMBERS

OBJECTIVES:

To understand integers, rational numbers, real numbers, and absolute values.

** REVIEW

*(A) The natural numbers (Nat) = { 1,2,3,...}
*(B) The whole numbers (Wh) = { 0,1,2,3...}
*(C) The integer numbers (Int) = {...-2,-1,0,1,2,3,...}
*(D) The rational numbers (Rat) = {a/b : a and b are integers, b ≠ 0}
*(E) The irrational numbers (Irrat) = numbers which cannot be expressed by a rational number.
*(F) The real numbers (Real) = all **rational** and **irrational** numbers.

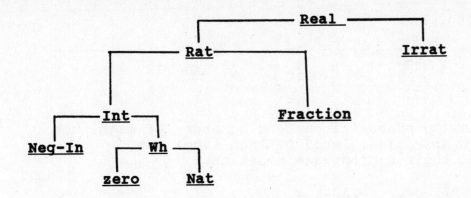

Reals consist of both rational and irrational numbers.
Rationals consist of both integers and fractions.
Integers consist of negative integers and whole numbers.
Whole numbers consist of the natural numbers and zero.

Example:

	__Nat__	__Wh__	__Int__	__Rat__	__Irrat__	__Real__	__None__
5	y	y	y	y	n	y	n
-3	n	n	y	y	n	y	n
3/8	n	n	n	y	n	y	n
-4/17	n	n	n	y	n	y	n
0	n	y	y	y	n	y	n
√7	n	n	n	n	y	y	n
7.3	n	n	n	y	n	y	n
0.123	n	n	n	y	n	y	n
√16	y	y	y	y	n	y	n
2π	n	n	n	n	y	y	n
√−3	n	n	n	n	n	n	y

Note: π is approximately equal to 3.1415927.

• Study the number line carefully.
The numbers to the **left** of zero are negative; the numbers
to the **right** of zero are positive.

If a number **x** lies to the right of another number **y** on the number
line, we say **x** is greater than **y** (or equivalently, **y** is less than **x**).

5

The number line:

```
    —.___.___.___.___.___.___.___.___.___.___.___.___._____
...  -6   -5   -4   -3   -2   -1    0   +1   +2   +3   +4   +5   +6 ...
```

**EXERCISES

1. Arrange the following numbers in ascending order of magnitude
   (**Hint**: Draw the number line. Label 0. Then label the
   numbers at their appropriate positions.)

   $$-10, \quad 5, \quad -2, \quad 2, \quad -5/12, \quad 0$$

2. Determine the absolute values of the following numbers:
   (Hint: The absolute value is the value **regardless** of the **sign**)

   a. $|-4| = ?$ b. $|+3| = ?$ c. $|-(-5)| = ?$ d. $|(-3) \cdot 14| = ?$

   e. $|-4/7| = ?$ f. $|-4| \cdot |-3| = ?$ g. $|(-4) \cdot (-3)| = ?$

   h. $|-4| \cdot |0| = ?$

3. Classify the number 12/4 as real, integer, rational, irrational,
   whole number, and/or natural number.

♦♦♦♦♦

**SOLUTIONS TO EXERCISES

1. Arrange the following numbers in ascending order of magnitude.

   $$-10 < -2 < -5/12 < 0 < 2 < 5$$

```
   —._____._____.__.___._____._____
   -10                         -2  -5/12  0   2          5
```

2. Determine the absolute values of the following numbers:

   a. $|-4| = 4$ b. $|+3| = 3$ c. $|-(-5)| = 5$

   d. $|(-3) \cdot 14| = |-42| = 42$ e. $|-4/7| = 4/7$

   f. $|-4| \cdot |-3| = 4 \cdot 3 = 12$ g. $|(-4) \cdot (-3)| = |12| = 12$

   h. $|-4| \cdot |0| = 4 \cdot 0 = 0$

3. Classify the number 12/4

   $12/4 = 3$ (Real, Rational, Integer, Whole number
                and Natural number)

♦♦♦♦♦

6

1.3 ADDITION and SUBTRACTION of REAL NUMBERS

OBJECTIVES:

*(A) To add two real numbers having <u>like signs</u> (the same sign).
*(B) To add two real numbers having <u>opposite signs</u>.
*(C) To subtract two real numbers having <u>like signs</u>.
*(D) To subtract two real numbers having <u>opposite signs</u>.

** REVIEW

*(A) Add two real numbers having like signs.
  Add their absolute values and keep the common sign.

Examples:
    a. $4 + 5 = ?$ b. $(-8) + (-6) = ?$

Solutions:
    a. $4 + 5$ --> the common sign is +
       $4 + 5 = + (|4| + |5|) = 9$

    b. $(-8) + (-6)$ ----> the common sign is -
       Keep the common sign and find the absolute value of each number.
       (The absolute value is the value **regardless** of **sign**).

         $(-8) + (-6) = - (|-8| + |-6|) = -(8 + 6) = -14$

*(B) Add two real numbers having opposite signs.
  Find the difference between their absolute values and keep the
  the sign of the number with the larger absolute value.

Examples:
    a. $-19 + 8 = ?$ b. $25 + (-16) = ?$ c. $2 + (-12) = ?$

Solutions:
    a. Find the difference of the absolute values of these two numbers.
       $|-19| - |8| = 19 - 8 = 11$
       Keep the sign of the number with the larger absolute value,
       which is -19. Therefore, the solution is -11.
       $-19 + 8 = -11$

    b. $25 + (-16) = +(|25| - |-16|) = +(25 - 16) = +9 = 9$

    c. $2 + (-12) = -10$

*(C) Subtract two real numbers having like signs.
  Change the sign of the second number and add the two real numbers.

Examples:
    a. $3 - 4 = ?$ b. $(-10) - (-5) = ?$

Solutions:
    a. 3 - 4   Change the sign of the second number which is 4 to (-4)
              Add the two real numbers
       3 - 4 = 3 + (-4) = -1

    b. (-10) - (-5) =(-10) + (+5) = -5
       Change -5 to +5 and then add.

*(D) Subtract two real numbers having opposite signs.
     Perform **exactly** the same steps as (C) above.

   Examples:
     a. 9 - (-5) = ?   b. (-12) - (9) = ?

Solutions:
     a. 9 - (-5) = ?   --> Change -5 to +5 and add the two
        9 - (-5) = 9 + (+5) = 14   real numbers

     b. (-12) - (9) = (-12) + (-9)   --> Add the two reals having like
                                          signs. The common sign is -
        -(|-12| + |-9|) = -(12 + 9) = -21

♦♦♦♦♦

EXERCISES

Perform the following operations .
     1. (-5) + (-4) = ?   2. (-5) + (4) = ?   3. (0) - (5) = ?
     4. 5 - (0) = ?   5. (0) - (-5) = ?   6. 0 + (-5) = ?
     7. (-12) - (-10) = ?   8. (-5) + 7 + (-4) = ?
     9. -120 + 120 - (-45) = ?
    10. 0 - 5 - 6 - 7 = ?   Hint:(0 - 5) - 6 - 7 = [(-5 + -6)] + (-7)
    11. 99 - (- 5) = ?
    12. -(-7) - (-7) = ?

♦♦♦♦♦

SOLUTIONS TO EXERCISES

<u>Hint</u>: Refer to (A) for problem 1,10.
       Refer to (B) for problem 2,6,8.9
       Refer to (C) for problems 3,4,7,10.
       Refer to (D) for problem 5,9,11,12.

1. (-5) + (-4)= -(|-5| + |-4|) = -(5 + 4) = -9   (The common sign is -)
2. (-5) + (4) = -(|-5| - |4|) = -(5 - 4) = -1
   Keep the sign of the number with the larger absolute value
   (which is -) and then subtract
3. (0) - 5 = 0 + (-5) = - 5
4. 5 - 0 = 5
5. 0 - (-5) = 0 + (+5) = 5
6. 0 + (-5) = -(|-5| - |0|) = -(5 - 0) = -5
7. (-12)-(-10) = (-12) + (+10) = -(|-12| - |10|) = -(12 -10) = -2

8

8. $(-5) + 7 + (-4) = [(-5) + 7] + (-4) = [(|7| - |-5|)] + (-4)$
   $= (7 - 5) + (-4) = (2) + (-4) = -2$
9. $-120 + 120 - (-45) = 0 - (-45) = 45$
10. $0 - 5 - 6 - 7 = -(|5| + |6| + |7|) = -18$ (The common sign is -)
11. $99 - (-5) = 99 + 5 = 104$
12. $-(-7) - (-7) = [-(-7) + 7] = 7 + 7 = 14$ $(-(-7) = +7)$

◆◆◆◆◆

1.4 <u>MULTIPLICATION and DIVISION of REAL NUMBERS</u>

<u>OBJECTIVES</u>:

*(A) To multiply two real numbers
*(B) To divide two real numbers

** <u>REVIEW</u>

*(A) Multiplying two real numbers having like signs.
    Dividing two real numbers having like signs.
    The result is always positive (+).

Examples:
     a. $8 \cdot 7 = ?$ b. $(-8) \cdot (-7) = ?$
     c. $24/8 = ?$ d. $(-24)/(-4) = ?$

Solutions:
     a. $8 \cdot 7 = 56$ b. $(-8) \cdot (-7) = 56$ c. $24/8 = 3$
     d. $(-24)/(-4) = 6$

*(B) Multiplying two real numbers having opposite signs.
    Dividing two real numbers having opposite signs.
    The result is always negative (-)

Examples:
     a. $(-8) \cdot 7 = ?$ b. $(-24)/4 = ?$ c. $24/(-8) = ?$

Solutions:
     a. $(-8) \cdot 7 = -56$ b. $(-24)/4 = -6$ c. $24/(-8) = -3$

◆◆◆◆◆

** <u>EXERCISES</u>

Evaluate the following:

1. $6 \cdot 8 = ?$ 2. $(-0.5) \cdot 20 = ?$ 3. $(-3) \cdot (-6) = ?$
4. $5 \cdot (-0.25) = ?$ 5. $10/4 = ?$ 6. $10/(-4) = ?$
7. $(-0.5)/(-5) = ?$ 8. $(-6)/12 = ?$ 9. $(-1.25) \cdot (-4) = ?$
10. $(1000.0/10.0) = ?$

◆◆◆◆◆

1. $6 \cdot 8 = 48$
2. $(-0.5) \cdot 20 = -(0.5 \cdot 20) = -10$
3. $(-3) \cdot (-6) = 3 \cdot 6 = 18$
4. $5 \cdot (-0.25) = -1.25$
5. $10/4 = 2.5$
6. $10/(-4) = -(10/4) = -2.5$
7. $(-0.5)/(-5) = +(0.5/5) = 0.1$
8. $(-6)/12 = -(6/12) = -1/2 = -0.5$
9. $(- 1.25) \cdot (-4) = 5$
10. $(1000.0/10.0) = 100$

♦♦♦♦♦

1.5 ARITHMETIC EXPRESSIONS

OBJECTIVES:

*(A) To manipulate exponential expressions
*(B) To understand the precedence of the mathematical operators

REVIEW

*(A) Manipulation of exponential expressions
The notation a^k means a times itself k times,
where a is the base and k is the exponent or power.
Therefore $a \cdot a \cdot a \cdots a$ k times is denoted by a^k.
Examples:
Write each product in exponential notation.
Write each exponential expression as a product without exponents.

a. $-3^2 = -(3 \cdot 3) = -9$
b. $(-3)^2 = (-3) \cdot (-3) = 9$
c. $x^4 = x \cdot x \cdot x \cdot x$
d. $(-3)(-3)(-3)(-3)(-3) = (-3)^5$
e. $(1.2)(1.2)(1.2) = (1.2)^3$

*(B) Precedence of the mathematical operators.
Evaluate the expression according to the following order of precedence:

exponentiation	↑	highest priority
x , ÷		
+ , -		lowest priority

If the operators are at the same level of priority and they are next to each other, then evaluate from left to right.

Examples:

a. $3^2 \cdot (-2)^3 = (3 \cdot 3) \cdot [(-2)(-2)(-2)] = 9 \cdot (-8) = -72$
b. $4 \cdot 3 \div 2 + 10 - 5 = [(4 \cdot 3) \div 2] + 10 - 5 =$
   $[12 \div 2] + 10 - 5 = 6 + 10 - 5 = 16 - 5 = 11$

♦♦♦♦♦

EXERCISES

I. Write each product as an exponential expression.
   Write each exponential expression as a product without exponents.

10

1. $-(-2)^3 = ?$ 2. $(a)(a)(a)(a) = ?$ 3. $(-b)^3 = ?$
4. $(-2)(-2)(-2)(-2)(-2)(-2) = ?$ 5. $-(b)(b)(b) = ?$

II. Evaluate each of the following exponential expressions:

6. $(10)^5$ 7. $(-4)^5$ 8. 6^4
9. $-(|-2|)^5$

III. Evaluate each of the following expressions:

10. $(-3)^2(-2) + (5)^2$ 11. $(9 - 9^0)/(10^2 - 2^2)$ 12. $(8)^2 - 2^2$

13. $((-2)^2(-1)^3)^2$ 14. $|3-2^2| + |2^4-3|$ 15. $(-12)^2 - (4.0)$

16. $-(-|-32|)$ 17. $6^3 + (-6)^3$

18. $(-2)[9^2-(-5)^0(-2)^3]$

19. $12 - |3\cdot(-22)|$ 20. $(-5)^5+|-3.125|$

♦♦♦♦♦

SOLUTIONS TO EXERCISES

I. Write each product as an exponential expression.
   Write each exponential expression as a product without exponents.

1. $-(-2)^3 = -[(-2)(-2)(-2)] = -(-8) = 8$

2. $(a)(a)(a)(a) = a^4$ 3. $(-b)^3 = (-b)(-b)(-b)$

4. $(-2)(-2)(-2)(-2)(-2)(-2) = (-2)^6$ 5. $-(b)(b)(b) = -b^3$

II. Evaluate each exponential expression.

6. $(10)^5 = 100000$ 7. $(-4)^5 = -1024$

8. $6^4 = 1296$ 9. $-(|-2|)^5 = -(2)^5 = -32$

III. Evaluate each expression

10. $(-3)^2(-2) + (5)^2 = 9\cdot(-2) + 25 = -18 + 25 = 7$

11. $9-9^0 = 9-1 = 8$ and $10^2-2^2 = 100-4 = 96$
    Therefore, $(9-9^0)/(10^2-2^2) = 8/96 = 1/12$

12. $(8)^2 - 2^2 = 64 - 4 = 60$

13. $((-2)^2(-1)^3)^2 = (-4)^2 = 16$

14. $|3-2^2| + |2^4-3| = |-1| + |13| = 14$

15. $(-12)^2 - (4.0) = 140$ 16. $-(-|-32|) = 32$

11

17. $6^3 + (-6)^3 = 0$ 18. $(-2)[81-1(-8)] = -2 \cdot 89 = -178$

19. $12 - |3 \cdot (-22)| = -54$

20. $(-5)^5 + |-3.125| = -3121.875$

♦♦♦♦♦

1.6 <u>ALGEBRAIC EXPRESSIONS</u>

<u>OBJECTIVES</u>:

*(A) To understand algebraic expressions
*(B) To evaluate algebraic expressions
*(C) To solve algebraic equations

**<u>REVIEW</u>

*(A) An algebraic expression is a combination of letters (variables) and
      numbers connected by operational symbols.

  Examples:

      $3a^3 + b^2$, $x^5 - (4xy + 10)$, $(a - b^2)/(3a \cdot 5b)$

*(B) Evaluate algebraic expressions.

  Examples:
    Evaluate the following expressions given that $a = 1$, $b = -2$, $c = 0$,
    and $d = 3$.

    a. $2a^2 - 4ac = 2(1)^2 - 4(1) \cdot (0) = 2$

    b. $(d^3 - a \cdot b)/(3c + 4d) = [27 - (1) \cdot (-2)]/[(3 \cdot 0 + 4 \cdot 3)] = 29/12$

*(C) Solve algebraic equations.

  Example:
        $8x - 4 = 0$
        $(8x - 4) + 4 = 0 + 4$ add 4 on both sides
        $8x = 4$ simplify
        $x = \frac{1}{2}$ $\frac{1}{2}$ is the solution to the equation.

♦♦♦♦♦

**<u>EXERCISES</u>

I. Evaluate the following expressions given that $a = 2$, $b = -3$, $c = -2$.

    1. $-(4a + 5b)$ 2. $(-a + b)(a + b)$ 3. $c^3 - a^3$

    4. $a - b - c$ 5. $([|b - 5a| + 12])/(5c)$ 6. $a \div c \cdot b$

12

7. $c^5 - 12a$ 8. $|-c + (3b) \cdot (5c)|$

II. Solve each of the following equations.

9. $x - 12 = 0$ 10. $2x + 3 = 3$ 11. $\frac{1}{4}x - \frac{1}{4} = 0$
12. $\frac{3}{4} + x = \frac{3}{4} + 3$ 13. $x/9 = 1/9$ 14. $(21-2x)/7 = 3$
15. $-3x-5x = 2x + 4x$ 16. $8 = x/8 - 6$ 17. $-7x + 3x = 12$
18. $x/999 = 1$ 19. $3x + x/11 = 2$
20. $1x + 2x + 3x + 4x + 5x = 15$

♦♦♦♦♦

SOLUTIONS TO EXERCISES

I. Evaluate each expression given that $a = 2$, $b = -3$, $c = -2$.

1. $-(4a + 5b) = -(4 \cdot 2 + 5 \cdot -3) = -(8 - 15) = -(-7) = 7$
2. $(-a + b)(a + b) = (-2 + (-3))(2 + (-3)) = (-5)(-1) = 5$
3. $c^3 - a^3 = (-2)^3 - (2)^3 = -8 -8 = -16$
4. $a - b - c = 2 - (-3) - (-2) = 2 + 3 + 2 = 7$
5. $([|b - 5a| + 12])/(5c) = [13 + 12]/(-10) = -2.5$
6. $a \div c \cdot b = 2 \div (-2) \cdot (-3) = (-1) \cdot (-3) = 3$
7. $c^5 - 12a = (-2)^5 - 12(2) = -32 - 24 = -56$
8. $|-c + (3b) \cdot (5c)| = |-(-2) + (3(-3)) \cdot (5(-2))| = 92$

II. Solve each of the following equations.

9. $x - 12 = 0$ 10. $2x + 3 = 3$ 11. $\frac{1}{4}x - \frac{1}{4} = 0$
   $x = 12$ $x = 0$ $x = 1$

12. $\frac{3}{4} + x = \frac{3}{4} + 3$ 13. $x/9 = 1/9$ 14. $(21-2x)/7 = 3$
    $x = 3$ $x = 1$ $x = 0$

15. $-3x-5x = 2x + 4x$ 16. $8 = x/8 - 6$ 17. $-7x + 3x = 12$
    $x = 0$ $x = 112$ $x = -3$

18. $x/999 = 1$ 19. $3x + x/11 = 2$
    $x = 999$ $x = 11/17$

♦♦♦♦♦

20. $1x + 2x + 3x + 4x + 5x = 15$
    $x = 1$

1.7 **PROPERTIES of the REAL NUMBERS**

OBJECTIVES:

*(A) The commutative property
*(B) The associative property
*(C) The distributive property
*(D) The identity property
*(E) The inverse property
*(F) The multiplication property of zero

**<u>REVIEW</u>

*(A) The commutative property:
  - The <u>order</u> of addition of two real numbers <u>**does**</u> <u>**not**</u> <u>**affect**</u> the result.
  - The <u>order</u> of multiplication of two real numbers <u>**does**</u> <u>**not**</u> <u>**affect**</u> the result.
  - Subtraction and division do not follow the commutative property.

Examples:
    Commutative under + and ·
    $8 + 5 = 5 + 8$ and $24 \cdot 6 = 6 \cdot 24$
    Commutative property does not apply to subtraction and division
    $8 - 5 \neq 5 - 8$ and $24 \div 6 \neq 6 \div 24$

*(B) The associative property:
  - The order of the grouping in addition does not affect the result.
  - The order of the grouping in multiplication does not affect the result.
    $(12 + 5) + 3 = 12 + (5 + 3)$ and $(60 \cdot 6) \cdot 2 = 60 \cdot (6 \cdot 2)$
    but
    $(12 - 5) - 3 \neq 12 - (5 - 3)$ and $(60 \div 6) \div 2 \neq 60 \div (6 \div 2)$

*(C) The distributive property
  - The product of a real number and the sum (or difference) of two real numbers can be distributed over each term in the sum (or difference).

Examples:

    $3 \cdot (a + b) = 3 \cdot a + 3 \cdot b = 3a + 3b$ and $3 \cdot (a - b) = 3 \cdot a - 3 \cdot b$

*(D) The additive identity is 0 and the multiplicative identity is 1.

Examples:
    $33 + 0 = 0 + 33 = 33$ and $33 \cdot 1 = 1 \cdot 33 = 33$
    0 added to any real number does not change the value of the number.
    1 times any number does not change the value of the number.

*(E) The inverse properties:
    The additive inverse of any real number a is -a.
    The multiplicative inverse of any nonzero number a is 1/a.

Examples:
    The additive inverses of -0.5, -1/7, and -(-3)
    are 0.5, 1/7, and -3, respectively.
    The multiplicative inverse of -1/0.3, -7, and 1/3
    are -0.3, -1/7, and 3, respectively.

*(F) Multiplication Property of Zero.
    Any number multiplied by a zero is zero.

Examples:
     6·0 = 0   (-598)·0 = 0

♦♦♦♦♦

**EXERCISES

I. Identify the property that justifies each equality.

   1. 3·(1/3) = 1   2. (-1/3) + (1/3) = 0
   3. 9 + 0 = 0 + 9 = 9   4. 3·(5 + a) = 3·5 + 3·a = 15 + 3a
   5. (-1/3) + (1/3) = 0   6. (1/5)·1 = 1·(1/5) = 1/5
   7. 45·0 = 0   8. 3 + (b + c) = 3 + (c + b)

II. Find the additive and multiplicative inverses of each of the
    following:
                    additive inverse   multiplicative inverse
     9. ½   _____   _____
    10. -3   _____   _____
    11. 1/8   _____   _____
    12. -0.173   _____   _____

III. Complete each statement using the property named:

   13. 3 + 5 = _____, commutative   14. 8a + 8b = _____, distributive
   15. 3·(a·b) = _____, associative 16. -3(8 - y) _____, distributive
   17. -6(1) = _____, identity   18. (-3) + (_____) = 0, inverse
   19. 34·(_____) = 0, zero   20. -(-3) + _____ = 0, inverse

♦♦♦♦♦

**SOLUTIONS TO EXERCISES

I. Property that justifies each equality.

   1. 3·(1/3) = 1   Multiplicative inverse
   2. (-1/3) + (1/3) = 0   Additive inverse
   3. 9 + 0 = 0 + 9 = 9   Commutative and identity
   4. 3·(5 + a) = 3·5 + 3·a = 15 + 3a   Distributive
   5. (-1/3) + (1/3) = 0   Additive inverse
   6. (1/5)·1 = 1·(1/5) = 1/5   Commutative and identity
   7. 45·0 = 0   Multiplication of zero
   8. 3 + (b + c) = 3 + (c + b)   Commutative

II. Additive and multiplicative inverses:

		additive inverse	multiplicative inverse
9.	½	-½	2
10.	-3	3	-1/3
11.	1/8	-1/8	8
12.	-0.173	0.173	-(1/0.173)

15

III. Statements using the properties named:

13. $3 + 5 = \underline{5 + 3}$ commutative   14. $8a + 8b = \underline{8(a + b)}$ distributive
15. $3 \cdot (a \cdot b) = \underline{(3 \cdot a) \cdot b}$ associative
16. $-3(8 - y) = \underline{(-3) \cdot 8 - (-3) \cdot y} = -24 + 3y$ distributive
17. $-6(1) = \underline{-6}$ identity   18. $(-3) + \underline{3} = 0$ inverse
19. $34 \cdot \underline{0} = 0$ zero   20. $-(-3) + \underline{(-3)} = 0$ inverse

◆◆◆◆◆

1.8 **USING the PROPERTIES**

OBJECTIVES:

*(A) To perform sums and differences of algebraic expressions
*(B) To perform products and quotients of algebraic expressions
*(C) To remove parentheses

** **REVIEW**

*(A) Terms containing the same variables with the same exponents are called <u>like terms</u>.

Examples:
     $3a^2bc$, $5a^2bc$, $-10a^2bc$, and $\frac{1}{4}a^2bc$ are like terms

     $-6ab^4c$, $10a^2bc$, $8abc$, and $-1/7abc^7$ are unlike terms

● To simplify an algebraic expression involving additions and subtractions, combine like terms and use the distributive property.

Example:
    $5a - 3xy + 9a^5 + 7xy - 2a^5 - 2xy + 10a$ (Combining like terms)
    $= (5a + 10a) + (-3xy + 7xy - 2xy) + (9a^5 - 2a^5)$
    $= (5 + 10)a + (-3 + 7 - 2)xy + (9 - 2)a^5$ (Distributive)
    $= 15a + 2xy + 7a^5$

*(B) To simplify the product or quotient of an algebraic expression using one or more of the following properties:
commutative, associative, identity, and inverse properties

Examples:
    a. $(3a)(6b) = (3a \cdot 6) \cdot b = ((3 \cdot 6) \cdot a) \cdot b = 18ab$
    b. $(-5x)(y/5) = (-5x) \cdot [(1/5) \cdot (y)] = [-5 \cdot (1/5)] \cdot [(x \cdot y)] = -xy$
    Remember that: $[-(a + b - c)] = (-1) \cdot (a + b - c) = -a - b + c$

◆◆◆◆◆

** **EXERCISES**

I. Simplify the following algebraic expressions:

16

1. $(-3a + 5) + (-4 + 3a)$ 2. $3x + x(a + 8)$ 3. $8a \cdot (3 + 2)$
4. $3a - 4b + 2(5b - 2a)$ 5. $-2(0.5 \cdot 10 - \frac{1}{2} \cdot 2)$ 6. $2\pi + 8\pi$
7. $2b + 5a - (-5b) - (-5a)$ 8. $4m \cdot \frac{1}{4}m - 4$ 9. $(-9x + 15)/3$
10. $p-(q-r+p)$ 11. $3a - 3 + 2(a - 2) - 7$

II. True or False ?

12. $6(a - 3) = 6a - 18$ 13. $-3x - 9 = -3 \cdot (x - 3)$
14. $3a \cdot 2b = 6ab$ 15. $12a - 5b = 7ab$
16. $-(x - 4) = -x + 4$ 17. $x^2 - x = x$
18. $0.2(2a + 8a) = 0.2(10a) = 2a$ 19. $3 \cdot (-2x - 3x) = 18x^2$
20. $3 \cdot (-2x-3x) = 3 \cdot (5x) = 15x$

◆◆◆◆◆

**SOLUTIONS TO EXERCISES

I. Simplify:

1. $(-3a + 5) + (-4 + 3a) = 1$ 2. $3x + x(a + 8) = 11x + ax$
3. $8a \cdot (3 + 2) = 40a$ 4. $3a - 4b + 2(5b - 2a) = -a + 6b$
5. $-2(0.5 \cdot 10 - \frac{1}{2} \cdot 2) = -8$ 6. $2\pi + 8\pi = 10\pi$
7. $2b + 5a - (-5b) - (-5a) = 10a + 7b$
8. $4m \cdot \frac{1}{4}m - 4 = m^2 - 4$
9. $(-9x + 15)/3 = -3x + 5$
10. $p-(q-r+p) = r-q$
11. $3a - 3 + 2(a - 2) - 7 = 5a$

II. True/False

12. True.   13. False. $(-3x-9) = -3(x + 3)$ 14. True.
15. False.   16. True.
17. False. $(x^2-x = x(x-1))$ 18. True
19. False $(3 \cdot (-2x-3x) = -15x)$ 20. False $(-15x)$

◆◆◆◆◆

CHAPTER I - TEST A

I Fill-in.

	Natural	Whole	Integer	Rational	Real
1. -32					
2. $\sqrt{-3}$					
3. -3π					
4. 0.34567					
5. $\frac{1}{4}$					

II. Evaluate and simplify the following:

6. $|-3|-(|-8-5|)$ 7. $(-3/2) \cdot (8/9)$ 8. $12 \div (1/3)$
9. $(-0.005) \cdot (2000)$ 10. $3 \cdot 198 + 3 \cdot 2$ 11. $12 \cdot (a/4)$
12. $(18a + 4)/2$ 13. $a + 10 - 0.01(a + 100)$
14. $(3/7) \cdot (14/6)$ 15. $(3/17) \div (2/34)$ 16. $4 \cdot (\frac{1}{4} + 3\frac{1}{2})$

III. Evaluate each expression if $a = -1$, $b = 2$, and $c = 3$.

17. $a^2 - 2bc$ 18. $(3c-2b)/5a$ 19. $a^6 + c^3 - 6c$ 20. $b^5 - ac$

◆◆◆◆◆

CHAPTER I - TEST B

I. Fill in the missing values.

1. $30/50 = ?/5 = 15/? = 6/10 = 21/? = ?/150 = 180/? = ?/20$
2. $0.36 = ?/? = ?\%$ 3. $3/5 = ?\% = ?$ (in decimal)

II. Evaluate and simplify the following:

4. $3/7 - 14/6$ 5. $9/10 \div 18$ 6. $(4/12)(-12/4)$
7. $-|16+(-3)| \cdot 5$ 8. $|-(3 \cdot 5+2)|$ 9. $43 \cdot 3 - 43 \cdot 2$
10. $0 \cdot 3425$ 11. $3a + 5(3-2a)-3(2a-3)$
12. $[(a-b)/(b-c)] \cdot [(b-c)/(a-b)]$ 13. $1 \div (\frac{1}{2}) \div (1/3) \div (\frac{1}{4})$
14. $3a + [8a \cdot (1/8)a]$

III. Evaluate each of the following expressions if $a=5$, $b=0$, $c=-3$:

15. $b^2 - 4ac$ 16. $a \cdot a + a \cdot b + a \cdot c$ 17. $a^2 - b^2 - c^2$

IV. Solve each of the following equations:

18. $-3x + 4 = 2x - 1$ 19. $-x = 3(x + 2)$
20. $3 + 2x - (3/5) \div (1/5) = 1$

◆◆◆◆◆

CHAPTER I - TEST C

I. Evaluate and/or simplify each of the following:

1. $49/7$ 2. $40/90$ 3. $8/80000$ 4. $(1/2) \cdot (1/3)$
5. $(1/3) \cdot (3/5)$ 6. $(1/5) \cdot 100$ 7. $(3/10) \div (6/2)$ 8. $9 \div (1/3)$
9. $12/15 - 10/15$ 10. $(1/20) - (3/20)$ 11. $|0|-|8 + (-2)|$
12. $\frac{1}{2}-(-1/5)$ 13. $0.3 - 7$ 14. $1-(\frac{1}{2})-(1/3)-(\frac{1}{4})$
15. $(-1/3) \cdot (15/20)$ 16. $(-5.0) \div (0.5)$ 17. $-2^3 \cdot (-2)^2$
18. $2 \cdot 3 - 3^3 + 9/3$ 19. $3 - 2 \cdot |8-(8^2-6^2)|$ 20. $8 - (x+5) = -1 + x$

◆◆◆◆◆

18

CHAPTER I - TEST A - SOLUTIONS

I Fill-in.

	Natural	Whole	Integer	Rational	Real
1. -32	n	n	y	y	y
2. $\sqrt{3}$	n	n	n	n	y
3. -3π	n	n	n	n	y
4. 0.34567	n	n	n	y	y
5. ¼	n	n	n	y	y

II Evaluate and simplify the expressions:

6. $|-3| - (|-8-5|) = 3 - 13 = -10$ 7. $(-3/2) \cdot (8/9) = -4/3$

8. $12 \div (1/3) = 36$ 9. $(-0.005) \cdot (2000) = -10$

10. $3 \cdot 198 + 3 \cdot 2 = 3 \cdot (198+2) = 600$ 11. $12 \cdot (a/4) = 3a$

12. $(18a + 4)/2 = 9a + 2$

13. $a + 10 - 0.01(a + 100) = a + 10 - 0.01a - 1 = 0.99a + 9$

14. $(3/7) \cdot (14/6) = 1$

15. $(3/17) \div (2/34) = 3$

16. $4 \cdot (¼ + 3½) = 4 \cdot (15/4) = 15$

III. Evaluate each expression (a = -1, b = 2, and c = 3):

17. $a^2 - 2bc = (-1)^2 - 2(2)(3) = 1-12 = -11$

18. $(3c-2b)/5a = (9-4)/(-5) = -1$

19. $a^6 + c^3 - 6c = (-1)^6 + 3^3 - 6(3) = 1 + 27 - 18 = 10$

20. $b^5 - ac = (2)^5 - (-1)(3) = 32 + 3 = 35$

◆◆◆◆◆

CHAPTER I - TEST B - SOLUTIONS

I. Fill-in:

1. $30/50 = \mathbf{3/5} = 15/\mathbf{25} = 6/10 = 21/\mathbf{35} = 90/\mathbf{150} = 180/\mathbf{300} = \mathbf{12}/20$

2. $0.36 = \mathbf{36/100} = \mathbf{36\%}$

3. $3/5 = \mathbf{0.6 = 60\%}$

II. Evaluate and simplify the expressions:

4. $3/7 - 14/6 = (18/42) - (98/42) = -(80/42) = -(40/21)$

5. $9/10 \div 18 = (9/10) \cdot (1/18) = 1/20$

6. $(4/12)(-12/4) = (4/12) \cdot (-12/4) = -1$

7. $-|16+(-3)| \cdot 5 = -13 \cdot 5 = -65$

8. $|-(3 \cdot 5+2)| = 17$

9. $43 \cdot 3 - 43 \cdot 2 = 43(3-2) = 43$

10. $0 \cdot 3425 = 0$

11. $3a + 5(3-2a) - 3(2a-3) = 3a + 15 - \underline{10a} - \underline{6a} + 9 = -13a + 24$

12. $[(a-b)/(b-c)] \cdot [(b-c)/(a-b)] = 1$
13. $1 \div (\frac{1}{2}) \div (1/3) \div (\frac{1}{4}) = (1 \cdot 2) \div (1/3) \div (\frac{1}{4}) = (2 \cdot 3) \div (\frac{1}{4}) = 6 \cdot 4 = 24$
14. $3a + [8a \cdot (1/8)a] = 3a + a^2$

III. Evaluate each expression ($a = 5$, $b = 0$, $c = -3$):

15. $b^2 - 4ac = (0) - 4(5)(-3) = 60$
16. $a \cdot a + a \cdot b + a \cdot c = a(a + b + c) = 5 \cdot 2 = 10$ distributive
17. $a^2 - b^2 - c^2 = (5)^2 - (0)^2 - (-3)^2 = 25 - 0 - 9 = 16$

IV. Solve each equation

18. $-3x + 4 = 2x - 1$
    $-3x - 2x = -1 - 4$
    $-5x = -5$
    $x = 1$

19. $-x = 3(x + 2)$
    $-x - 3x = 6$
    $-4x = 6$
    $x = -3/2$

20. $3 + 2x - (3/5) \div (1/5) = 1$
    $3 + 2x - (3/5) \times (5/1) = 1$
    $2x = 1 - 3 + 3$
    $x = 1/2$

◆◆◆◆◆

CHAPTER I - TEST C - SOLUTIONS

1. $49/7 = 7$

2. $40/90 = 4/9$

3. $8/80000 = 1/10000$

4. $(1/2) \cdot (1/3) = 1/6$

5. $(1/3) \cdot (3/5) = 1/5$

6. $(1/5) \cdot 100 = 20$

7. $(3/10) \div (6/2) = (3/10) \cdot (1/3) = 1/10$

8. $9 \div (1/3) = 9 \cdot 3 = 27$

9. $12/15 - 10/15 = 2/15$

10. $(1/20) - (3/20) = -1/10$

11. $|0| - |8 + (-2)| = 0 - 6 = -6$

12. $\frac{1}{2} - (-1/5) = \frac{1}{2} + (1/5) = 7/10$

13. $0.3 - 7 = -6.7$

14. $1 - (\frac{1}{2}) - (1/3) - (\frac{1}{4}) = (1/12)(12 - 6 - 4 - 3)$
    $= (1/12)(-1) = -1/12$

15. $(-1/3) \cdot (15/20) = -1/4$

16. $(-5.0) \div (0.5) = (-5.0) \div (\frac{1}{2}) =$
    $= (-5.0) \cdot (2) = -10$

17. $-2^3 \cdot (-2)^2 = -8 \cdot 4 = -32$

18. $2 \cdot 3 - 3^3 + 9/3 = 6 - 27 + 3 = -18$

19. $3 - 2 \cdot |8 - (8^2 - 6^2)| = 3 - 2 \cdot 20 = -37$

20. $8 - (x+5) = -1 + x, \quad x = 2$

◆◆◆◆◆

CHAPTER 2

LINEAR EQUATIONS AND INEQUALITIES

IN ONE VARIABLE

2.1 LINEAR EQUATIONS in ONE VARIABLE

OBJECTIVES:

*(A) To learn the Addition-Subtraction Property of Equality
*(B) To learn the Multiplication-Division Property of Equality

** REVIEW

*(A) **Adding** or **subtracting** a number from both sides of an equation does not affect the equality of the equation.

Examples:
       a. x - 8 = 10   b. 3x + 7 = 10 + 2x
Solutions:
       a. x - 8 = 10   --> add 8 to both sides
          (x - 8) + 8 = 10 + 8   --> isolate x on one side of the equation
          x = 18

       b. 3x + 7 = 10 + 2x   --> subtract 7 from both sides
          3x = 3 + 2x   --> subtract 2x from both sides
          x = 3

*(B) **Multiplying** or **Dividing** a nonzero number to both sides of an equation **does** **not** affect the equality of the equation.

Examples:
       a. x/5 = 6   --> multiply 5 on both sides
          (x/5)·5 = 6·5   --> 5 is the inverse of 1/5
          x = 30
       b. 5x - 7 = 8 - 2x   --> add 2x on both sides
          (5x - 7) + 2x = (8 - 2x) + 2x   --> isolate x on one side
          7x - 7 = 8   --> use associative and commutative properties
          7x = 15   --> add 7 on both sides
          x = 15/7   --> divide both sides by 7

Hints:

• Simplify the equation so that the x term is isolated on one side of the equation and the number is on the other side.

• How do you simplify the equation?
  Step1: Add or subtract any number to both sides of the equation
  Step2: Multiply or divide a **nonzero** number on both sides of the equation

◆◆◆◆◆

21

** EXERCISES

I. Solve each linear equation and check your answer.

1. -x + 8 = 2
2. (7x)/12 = -21
3. -6.8x = 66 - 0.2x
4. 3x + 5 = 5
5. -8x + 1 = 2 - 7x
6. -x = 2/3 + ¼
7. 1.5x = 3/2
8. ½·x = 6/7
9. -4x - 5 = -7
10. ¾·x - 6 = -9
11. 1.75x - 5 = 0.75x + 0.75
12. -2(3x + 5) = 2x + 6
13. 0.05x + 0.25 = 1.75
14. 46 - x - 2x = -3 + 4x
15. x/3 = 1/6
16. x/5 = 1 + x/6
17. 2.5x - 3.75 = 3.25 - x
18. -x/2 - x/3 - x/4 = 26/12
19. 88 = x + 88·(x/8)
20. 5(3x - 5) = (1/5)(7x + 11)

♦♦♦♦♦

** SOLUTIONS TO EXERCISES

1. -x = 2-8, x = 6
2. 7x = -252, x = -36
3. -6.6x = 66, x = -10
4. 3x = 0, x = 0
5. -x = 1, x = -1
6. x = -(11/12)
7. 3x = 3, x = 1
8. x = 12/7
9. -4x = -2, x = ½
10. 3x - 24 = -36, x = -4
11. x = 5.75
12. -6x - 10 = 2x + 6, -8x = 16, x = -2
13. 0.05x = 1.50, x = 30
14. 7x =49, x = 7
15. 6x = 3, x = ½
16. 6x = 30 + 5x, x = 30
17. 3.5x = 7, x = 2
18. -13x = 26, x = -2
19. 88 = x + 11x, 12x = 88
    x = 7(1/3)
20. 25(3x-5) = 7x + 11,
    68x = 136, x = 2

♦♦♦♦♦

2.2 MORE LINEAR EQUATIONS

OBJECTIVES:

*(A) To recognize identity and inconsistent equations.
*(B) To solve equations involving fractions and/or decimals.

** REVIEW

*(A) An identity equation is an equation equal to itself.

Examples:
  a. 3x - 5 = 3x - 5   b. 6x = 6x
    There are infinitely many solutions to the above equations.
    To satisfy the above equations, x can be any number.

• An inconsistent equation is an equation that has **no solution**.

Examples:
    a. 3x = 3x + 5   b. 7x = 0 + 7x - 10
      There are **no solutions** to the above equations.

• A conditional equation is an equation that has at least one real
  solution.

22

*(B)
   a. To solve equations involving fractions:

      1. Find LCD and multiply both sides by the LCD.
      2. Simplify the equation by combining like terms.

Example:
       $x/3 + 1 = x/5 - 1/3$ --> find the LCD = 15
       $15(x/3 + 1) = 15(x/5 - 1/3)$ --> multiply both sides by the LCD
       $5x + 15 = 3x - 5$ --> distributive property
       $2x = -20$ --> simplify
       $x = -10$

   b. To solve equations involving decimals:

      1. Find the largest number of decimal places appearing in the
         decimal numbers of the equation.
      2. Multiply both sides of the equation by 10, 100, 1000, etc.
         depending upon the largest number of decimal places in the
         equation.

Example:
   $0.4x + 2.005 = 1.005$
   $1000(0.4x + 2.005) = 1000(1.005)$
   (The largest number of decimal places in the equation is 3
   in the number 2.005).
   Therefore, multiply both sides by 1000.
   $400x + 2005 = 1005$ --> subtract 2005 from both sides
   $400x = -1000$ --> simplify
   $x = -2.5$

◆◆◆◆◆

** EXERCISES

I. Identify each as a conditional equation, inconsistent equation,
   or identity equation.

   1. $0 \cdot (3x+1) = 5x + 3$ 2. $x - x = 5x$
   3. $8x + 3 = 5$ 4. $3/x + 2/x = 6/x$, where $x \neq 0$
   5. $4x + 4 = 0$ 6. $3x^2 + 2x + 1 = 3x^2 + 2x + 1$
   7. $3x + 1 = 3x + 2$ 8. $3 - 2(5x + 3) = -7x - 3$
   9. $5x \div x = 4x$ 10. $0 \cdot x + 3 = x + 0$

II. Solve each equation.

   Hints: a. Find **LCD** and eliminate the **fraction**.
          b. **Multiply** the **decimal** by **10, 100, or 1000**, etc.
             to eliminate the decimals.

   11. $x/3 + x/4 = 1$ 12. $x/2 + 5 = 10$
   13. $(5x)/6 = \frac{1}{2}$ 14. $7x + 3 = 2x - 5$
   15. $x/5 - (2x/5) = 1/5$ 16. $0.3x - 0.05(x + 1) = 0.2$
   17. $0.25(x-5) = 0.05(x + 1)$ 18. $0.1x - x = 99$
   19. $5.25x - 4.25x = 1.25$ 20. $28.5x - 5.3x = 5.3 + 0.2x$

23

** <u>SOLUTIONS TO EXERCISES</u>

1. Conditional   2. Conditional   3. Conditional
4. conditional (after simplifying the equation we have 6x = 5x)
5. Conditional   6. Identity   7. Inconsistent
8. Identity   9. Conditional   10. Conditional
11. $x = 12/7$ 12. $x = 10$ 13. $x = 3/5$
14. $x = -(8/5)$ 15. $x = -1$
16. $x = 1$ (multiply both sides by 100)   17. $x = 6.5$
18. $x = -110$ (multiply both sides by 10)   19. $x = 1.25$
20. $x = 5.3/23 = 53/230$

◆◆◆◆◆

2.3 <u>FORMULAS</u>

<u>OBJECTIVES</u>:

* To solve for a variable in an equation.

** <u>REVIEW</u>

* (A) Solve for a variable by isolating the variable on one side of
the equation and the other terms (constant terms) on the other
side of the equation.

Example:
Solve the equation $3p + 8r = \frac{1}{4}(p + q + r)$ for p
$4(3p + 8r) = (p + q + r)$ --> multiply both sides by LCD = 4
$12p + 32r = p + q + r$ --> distributive property
$11p = q - 31r$ --> combine like terms and simplify
$p = (q - 31r)/11$

*(B) Find the value of a variable
If x represents the unknown variable and refer to *(A) above
to solve the equation.

Examples:
a. If the circumference of a circle is 20, find the radius.
The circumference of a circle is $2\pi \cdot r$, where r is the radius of the
circle.

b. The formula $s = s_0 + vt$ is used to measure the distance that a
particle travels.
(v: velocity, s_0: initial distance and t: time it travels)
Find the time the particle travels if the distance it travels is
100 miles, the initial distance is 40 miles, and the velocity is
20 miles/hour.

Solutions:
a. Let x represent the unknown variable (radius), then $2\pi \cdot x = 20$
$x = 10/\pi$

b. $s = 100$, $s_0 = 40$, $v = 20$, and $t = ?$
$s = s_0 + vt$
$100 = 40 + 20t$, $t = 3$ (hours)

24

** EXERCISES

I. Solve each formula for the specified variable.

1. $5s = x + (x + 1) + (x + 2)$ for x
2. $s = s_0 + v \cdot t$ for v
3. $q = p + prt$ for p
4. $s = s_0 + v \cdot t + \frac{1}{2}at^2$ for v
5. $a = \frac{1}{2}(u + p) \cdot h$ for h
6. $I = Prt$ for r
7. $3x + \frac{1}{2}y = z$ for y
8. $3/4(a+b) = 4/3(b + c)$ for b
9. $x + 4.75 = 0.5(y + 0.2x)$ for x
10. $-3(a - 2b) = \frac{1}{4}(3b - 2a)$ for a
11. $\frac{1}{2}(a -b) = 2b - 5a + c$ for a
12. $1 - \frac{1}{2}x - \frac{1}{4}y = -1$ for y
13. $y - x - z - w = 0$ for w
14. $3x^2 + 5y = 9x^2 - 5 + 6y$ for y

II. Solve each of the following problems.

15. The area of a rectangle is 30 square inches. The width is 15 inches. Find the length.

16. An ice box is 3 feet long, 5 feet wide, and 2 feet high. What is its volume ?

17. If the radius of a circle is 3 feet, what is the circumference of the circle ?

18. The area of a trapezoid is 32 square feet. If one base is 3 feet and the other base is 5 feet, find its height.

19. If you pay $200 in simple interest on a loan of $2000 for 4 years, what is the rate ?

20. Find the list price if there is a 20% discount and the sale price is $400.

♦♦♦♦♦

** SOLUTIONS TO EXERCISES

1. $x = (5s - 3)/3$
2. $v = (s - s_0)/t$
3. $q = p(1 + rt),\ p = q/(1 + rt)$
4. $s - (s_0 + \frac{1}{2}at^2) = v \cdot t$ and $v = (s - s_0 - \frac{1}{2}a \cdot t^2)/t$
5. $h = 2a/(u + p)$
6. $r = I/Pt$
7. $y = 2z - 6x$ (LCD = 2)
8. $b = (9a - 16c)/7$ (LCD = 12)
9. $x = (5/9)y - (95/18)$ multiply each side by 100.
10. $a = -(2.1)b$
11. $a = (5b + 2c)/11$
12. $y = 8 - 2x$
13. $w = y - x - z$
14. $y = -(6x^2 - 5) = 5 - 6x^2$ multiply each side by the LCD = 4
15. $30 = 15 \cdot L,\ \ L = 2$ inches
16. $V = 3 \cdot 5 \cdot 2 = 30$ volume = length $\cdot$ width $\cdot$ height
17. $C = 2\pi \cdot 3 = 6\pi$ feet

18. $(3 + 5) \cdot (h/2) = 32$, $8h = 64$, $h = 8$ feet
19. $r = 2.5\%$ ($I = Prt$ and $200 = 2000 \cdot 4 \cdot r$ solve for r)
20. $L = \$500$ (sale price = list price - list price x discount)
   ($400 = L - L \cdot (0.2)$, $400 = 0.8L$, therefore $L = \$500$)

◆◆◆◆◆

2.4 __ENGLISH TO ALGEBRA__

__OBJECTIVES__:

*(A) To become familiar with consecutive integers and pairs of numbers.
*(B) To use formulas to solve word problems.
*(C) To write meaningful equations

** __REVIEW__

*(A) • If x denotes an unknown integer, then the __consecutive__ integers
        are: x, x+1, x+2, x+3, ...
     • If y denotes an unknown odd integer, then the __consecutive__ __odd__
        integers are: y, y+2, y+4, y+6 ... each number is larger than the
        previous integer.
     • Any pair of numbers is related in some way.

  Examples:
     a. How will you represent two numbers such that one number is 3
        times the other ?
     b. The sum of two numbers is 50. How are two numbers related ?
  Solutions:
     a. If x represents the unknown number, then the other number is
        related to x by 3 times, that is 3·x.
     b. If x is one number, then the other number is 50 - x.

*(B) Using formulas
Example:
 Express interest in terms of time.
 The simple interest formula: $I = Prt$
 If principal is $200 and the rate is 20%, then using the formula
         $I = (200) \cdot (0.2) \cdot t$
         $I = 40t$

*(C) Writing equations
     Identify the variable (unknown) and write an equation to
     describe the problem.

Examples:
 a. The sum of three consecutive integers is 21. Find the integers.
Solution:
 Let x be one of the integers and let x+1 and x+2 be the other
 two integers.
     $x + (x + 1) + (x + 2) = 21$
     $3x + 3 = 21$
     $x = 6$, $x + 1 = 7$, $x + 2 = 8$
     The three integers are 6, 7, 8.

26

b. John has 35 coins, all in nickels and dimes, amounting to $2.75.
   How many nickels and dimes does he have ?

  Solution:
     If x is the number of nickels, then 35 - x is the numbers of dimes.
     1 nickel = $0.05 and 1 dime = $0.10
     (0.05)·x + (35 - x)·(0.10) = 2.75   --> multiply both sides by 100
     100[(0.05)·x + (35 - x)·(0.10)] = 100·(2.75)
     5x + 350 - 10x = 275
     -5x = -75
     x = 15 (nickels) and (35 - 15) = 20 (dimes)

check: There is a total 15 + 20 = **30 coins.**
        15·($0.05) + 20·($0.10)   --> total amounts to $2.75
        $0.75 + $2.00 = $2.75

♦♦♦♦♦
** EXERCISES

I. True or False ?

  1. Four consecutive odd integers are: x, x+1, x+2, x+3.
  2. The value in cents of x nickels, y dimes, and z quarters is
     0.05x + 0.1y + 0.25z
  3. If the difference of two numbers is 5, then the two numbers are
     related by x, 5 - x.
  4. If the width is 50 inches less than the length of a rectangle,
     then the area of the rectangle is represented by A = x·(x - 50)
  5. David is three times as old as John. If x represents David's age,
     then John's age can be represented by 3·x.

II. Find algebraic expressions for each of the following:

  6. The difference of a number and 5.
  7. The product of a number and 8.
  8. Three consecutive integers with a sum of -12.
  9. The area, if the base of a triangle is x and the height is x + 5.
 10. The principal, if the time is 2 years and the rate is 8%, and
     the interest is 5x. (I = Ptr)
 11. 20 is 10% of what number ?
 12. What percent of 20 is 100 ?
 13. The angle C, if in a triangle ABC and angle A is twice as large as
     angle B.   (A + B + C = 180^0)
 14. The fraction whose numerator is 8 more than twice the
     denominator.
 15. The perimeter of the rectangle is 35. The width is 5 less than
     half of the length.

♦♦♦♦♦

** SOLUTIONS TO EXERCISES

1. False. (x, x+2, x+4, x+6)   2. True.   3. False.(x,x-5)
4. True.   5. False.   (David's age is x, John's age is x/3)

27

6. x - 5   7. 8·x   8. x + (x+1) + (x+2) = -12
9. A = x(x+5)/2   10. 5x = 0.08·2P = 0.16P, P = (5x)/0.16
11. 20 = x·10%   12. x%·20 = 100
13. A+**2A**+C = 180^0 (angle B = 2A)   14. (2x+8)/x
15. 2[(**L/2 - 5**) + L] = 35 and L = 15

check:
    perimeter = 2(width + length), width = (15/2)-5 = 2.5.
    Therefore perimeter = 2(2.5+15) = 35.

♦♦♦♦♦

2.5 **VERBAL PROBLEMS**

OBJECTIVES:

*(A) To learn the general strategy for all word problems.
*(B) To solve various word problems.

** **REVIEW**

*(A) Step 1. Let x represent the unknown.
     Step 2. Express each statement in terms of x.
     Step 3. Solve the equation.

Example:
    The sum of two integers is 40, and one number is 4 times the
    other. Find the numbers.
Solution:
    If x is the unknown integer, then 4x is the other integer.
    x + 4x = 40
    x = 8, 4x = 32
    The two numbers are 8 and 32.

*(B) Various word problems.

Example:
    a. John is paid $10 for each day he travels and forfeits $2 for
       each day he rests. At the end of 30 days he nets $240. How many
       days did he rest ?
Solution:
    Let x denotes the number of days he rested,
    then (30 - x) will be the number of days he traveled.
    10·(30-x) - 2(x) = 240   --> $10 for (30-x) days John traveled
                                     and $2 for x days John rested

    300 - 10x - 2x = 240
    -12x = -60
    x = 5 days (rested)

    b. David bought a sofa for $300. He got a discount of 25%. What was
       the original price of the sofa ?

28

Solution:
     Let x be the original price of the sofa
     25%·x   --> 25% discount of the original price of the sofa
     x - 25%·x   --> The price John paid for the sofa which is $300
     x - 0.25x = 300
     100x - 25x = 30,000   --> multiply each side by 100
     75x = 30,000 and x = $400

♦♦♦♦♦

** EXERCISES

Find an algebraic expression for each of the following:

1. The area of a triangle, if height is twice as much as base.
2. The sum of 5 times a number and ten is four less than the product of nine and the number.
3. The difference between five times a number and eight is thirty-four.
4. The sum of two odd consecutive numbers is 16.
5. Five more than three times an unknown number.
6. The difference between two numbers is 12 and their sum is 2.
7. The square of an integer.
8. The sum of the squares of two consecutive integers.
9. One third of an integer is 20 more than one fifth of the number.
10. Ten less than three times a certain number.
11. The difference between 30 and the product of 4 and a number equals 30.
12. The total of three times a number and the sum of the number and four equals twenty.

♦♦♦♦♦

** SOLUTIONS TO EXERCISES

1. $A = \frac{1}{2}(2x \cdot x)$ where x represents base
2. $5 \cdot x + 10 = 9 \cdot x - 4$
3. $(5 \cdot x - 8) = 34$
4. $x + (x + 2) = 16$
5. $5 + 3 \cdot x$
6. $(x - (2 - x)) = 12$
7. x^2
8. $x^2 + (x + 1)^2$
9. $(x/3) = (x/5) + 20$
10. $3x - 10$
11. $30 - 4x = 30$
12. $3x + (x + 4) = 20$

♦♦♦♦♦

2.6 INEQUALITIES

OBJECTIVES:

*(A) To learn basic inequalities.
*(B) To graph inequalities.
*(C) To write inequalities.

** REVIEW

*(A) To understand the basic concept of inequalities.

 - An inequality is a statement that one algebgaic expression is **greater than** or **less than** another algebraic expression.

Examples:
  a. $3 \cdot (-5) < -2$ b. $8 \geq (6 + 2)$
  c. $10 - 5 > 0$ d. $3(-3) + 12 \leq 3$

*(B) Graph inequalities

Examples: Sketch the graph of each inequality on the number line.

  a. $x > -3$ (x is greater than -3)

  ```
  ─────────────────○━━━━━━━━━━━━━━━
      ... -5   -4   -3   -2   -1   0 ...
  ```

  b. $x \leq 10$ (x is less than or equal to 10)

  ```
  ━━━━━━━━━━━━━━━●─────────────────
      ...  6   7   8   9   10   11   12 ...
  ```

  c. $-2 \leq x < 5$ (x is between 5 and -2 but including -2)

  ```
  ─────────●━━━━━━━━━━━━━━○─────────
      ... -3   -2   -1   0   1   2   3   4   5   6 ...
  ```

*(C) Write inequalities

Examples:
  a. John made an 85 on the first test and 90 on the second. In order to get an A, his average of three tests must be between 90 and 100. What must John score on the third test ?

  b. The perimeter of a circle must not be larger than 20 inches. What must be the radius ?

Solutions:
  a. Let x represent John's third test score, then the average is (85 + 90 + x)/3. The average must be between 90 and 100.
     $90 \leq (85 + 90 + x)/3 \leq 100$
     $90 \leq (175 + x)/3 \leq 100$
     $270 \leq 175 + x \leq 300$
     $95 \leq x \leq 125$ --> John must make at least 95 on the third
                                test in order to get an A.

30

b. $2\pi \cdot r \leq 20$
   $r \leq 10/\pi$ --> The radius must be less than $10/\pi$.

♦♦♦♦♦

** EXERCISES

I. True or False ?

  1. $7 > 6 + (-3)$
  2. $|-3| \geq |3|$
  3. The number 0 satisfies the inequality $(x - 5) > -3$
  4. The number 2 satisfies the inequality $x - 10 < -1$
  5. $-9 < -8 < -6$
  6. $9 < 8 < 6$
  7. $4 \cdot (-2) + 10 < 6 \cdot 3 \cdot (-2) - 10$
  8. $-3 \cdot 5 + 20 \leq 2 \cdot 5 - 6 \leq (-10) \cdot (-5)$
  9. $|-12 + 9| < |-9 + 12|$
  10. -9 is the solution to the inequality $3x - 8x - 12 > 10$

II. Sketch the graph of each inequality on the number line.

  11. $x < 4$ 12. $x \geq 2$ 13. $-8 \leq x$
  14. $x \geq 8$ 15. $-40 < x < 40$ 16. $-3 < x < 0$
  17. $-5 \leq x \leq -1$ 18. $9 < x \leq 9$ 19. $-10 < x < 2$
  20. $10 \geq x > 2$

♦♦♦♦♦

** SOLUTIONS TO EXERCISES

I. True / False

  1. True.   2. True.
  3. False. $(-5 < -3)$ 4. True. $(2 - 10 = -8 < -1)$
  5. True.   6. False. $(6 < 8 < 9)$
  7. False. (2 is not less than -46) 8. False. $(4 \leq 5 \leq 50)$
  9. False. $(|-12 + 9| = |-9 + 12| = 3)$
  10. True. $(3 \cdot (-9) + 8 \cdot (-9) - 12 = 33 > 10)$

II. Sketch the graph of each inequality on the number line.

11. $x < 4$

```
━━━·━━━━·━━━━·━━━━○────────────
... 1      2      3      4      5 ...
```

12. $x \geq 2$

```
────·────·────●━━━·━━━·━━━━━━━━
... 0      1      2      3      4      5 ...
```

13. $-8 \leq x$

```
────·────●━━━·━━━━·━━━·━━━━━
... -9     -8    -7    -6     -5     -4 ...
```

14. $x \geq 8$

```
────·────·────●━━━·━━━·━━━━·━━━
... 6      7      8      9     10     11 ...
```

31

15. -40 < x < 40

```
—·——————o════·════·════·════·════·════·════·════·════o———·————
  -50   -40  -30    -20    -10     0    10    20    30    40   50    60
```

16. -3 < x < 0

```
          —·———·———o════·════·════o———·———·———
        ...  -4    -3    -2    -1     0     1     2 ...
```

17. -5 ≤ x ≤ -1

```
          ————·————●════════════════●————————
        ...  -6    -5    -4    -3    -2    -1 ...
```

18. none

19. -10 < x < 2

```
  ————————o════·════·════·════·════·════·════·════·════·════·════o———·———
  ... -11  -10  -9   -8   -7   -6   -5   -4   -3   -2   -1   0    1    2    3 ...
```

20. 2 < x ≤ 10

```
  ————————————o════·════·════·════·════·════·════·════●———·————·————
        ...  1    2    3    4    5    6    7    8    9   10   11   12 ...
```

♦♦♦♦♦

2.7 <u>SOLVING INEQUALITIES</u>

<u>OBJECTIVES</u>:

*(A) To understand the rules of inequalities.
*(B) To learn the applications of inequalities.

** <u>REVIEW</u>

*(A) Two rules for inequalities:

● Rule 1. The inequality sign <u>remains</u> the same, if:
  a. <u>**A positive number**</u> is **added, subtracted, multiplied,** or **divided** on each side of the inequality.
  b. <u>**A negative number**</u> is **added or subtracted** on each side of the inequality.

Example:
```
  6 > 4
  6 + 5 > 4 + 5          --> add a positive number to both sides
  6 - (-7) > 4 - (-7)    --> subtract a negative number from both sides
  6·(12) > 6·(12)        --> multiply both sides by a positive number
  6 ÷ 3  > 6 ÷ 3         --> divide each side by a positive number
  The inequality sign remains the same.
```

● Rule 2. The inequality sign <u>**reverses**</u> if a negative number is
          multiplied or divided on each side.
Example:
```
    8 > 4
    8·(-9) < 4·(-9)       --> multiply each side by a negative number
                              and reverse the inequality sign
    8 ÷ (-2) < 8 ÷ (-2)   --> divide each side by a negative number
                              and reverse the inequality sign
```

32

*(B) Applications of inequalities

Example:
James made a 90 on the first test and 85 on the second.
In order to get an A, the average of his two tests and the final exam must be between 92 and 100. What score must he make on the final in order to get an A ?

Solution:
Let x represent the final exam score
(90 + 85 + x)/3   --> average of all scores
92 ≤ [(90 + 85 + x)/3] ≤ 100   --> the average must be between
                                        92 and 100
273 ≤ 175 + x ≤ 300
98 ≤ x ≤ 125
James must get at least 98 on his final exam in order to get an A.

♦♦♦♦♦

** EXERCISES

I. True or False ?

1. The inequality 8x < -5 is equivalent to -5 > 8x
2. The inequality 5x < -9 is equivalent to -5x < 9
3. The inequality 2x - 10 > 0 is equivalent to x > 5
4. The statement " x is not less than 100 " is written as x ≥ 100
5. If x is the sale price of a sofa and the sale tax is 12%, then the total price of the sofa is x + 0.12·x.

II. Solve and graph each of the following inequalities.

    6. x - 3 > 2   7. 2x + 5 < -8   8. 3x < 9
    9. 2x + 5 > 5   10. 3 - ½x < 0   11. 2x - 4 ≤ 5x + 6
   12. 8 - 4(x - 3) ≤ 2(x + 5) - 5
   13. 0.42x + 30 ≥ 0.18x - 4.8   14. 8x - 5 ≥ -5x· (3/4)
   15. 10 ≥ (3 - 5x)/3 ≥ 2   16. -4 ≤ 8(x + 3)·¼ < 12
   17. 0 ≤ (18y - 4)/5 < (2y - 3)/4   18. -0.5 < 6x-0.3 < 0.5
   19. Each of the two equal sides of an isosceles triangle is 4 feet more than the base. The perimeter must be at least 50 feet. What is the range of values for the base ?

   20. John made a 68 on the midterm exam. In order to get a B, what range of scores on the final exam would put John's average between 80 and 89 inclusive. (The final exam counts as 2/3 and the midterm as 1/3 of the grade)

** SOLUTIONS TO EXERCISES

    1. True.   2. False. (-5x ≥ 9)   3. True.   4. True.
    5. True.   6. x > 5   7. x < -6.5   8. x < 3
    9. x > 0   10. x > 6   11. x ≥ -10/3   12. x ≥ 2.5
   13. x ≥ -145   14. x ≤ 20/47   15. -(27/5) ≤ x ≤ -(3/5)

33

16. $-5 \leq x < 3$ 17. $(2/9) \leq y < (1/62)$ 18. $-1/30 < x < 2/15$

19. If x is the the base, then the equal sides of the isosceles triangle are 4 + x. Therefore perimeter = x + 2·(4 + x) and this quantity must be at least 50 feet.
$x + 2 \cdot (4 + x) \geq 50$
$3x \geq 42$
$x \geq 14$ (feet)

20. If x represents the final exam score, then the expression $[68 \cdot (1/3)] + [(2/3) \cdot x]$ represents the range of scores that John must get in order to get a B.
$80 \leq [68 \cdot (1/3) + (2/3) \cdot x)] \leq 89$
$240 \leq 68 + 2x \leq 267$
$172 \leq 2x \leq 199$
$86 \leq x \leq 99.5$
John must score 86 or better.

◆◆◆◆◆

CHAPTER II - TEST A

I. Solve each equation

1. $5x - 7 = 3$
2. $17 - 3x = 5x + 1$
3. $\frac{1}{4}x = 8$
4. $3(2x - 40) = 0$
5. $0.53x + 0.12 = 0.47x$
6. $(3/5)x = (-5)/3$
7. $4x - 3x = 3x - 16$
8. $(1/5)x - 1/5 = (1/10)x$
9. $(3 - x)/(3 - x) = 1, x \neq 3$
10. $0.6x + 0.3(x - 1) = 0.9 - 0.3x$
11. $0.1x + 0.01x + 0.001x = 111$
12. $-\frac{1}{4}x = 40$
13. $3x + 7 = 3x + 6$
14. $3x - 5 = 3x - 5$
15. $(1/3)x + \frac{1}{4}x = (1/7)x$
16. $7x - 7x = 0$

II. Encode each of the following phrases or sentences as an algebraic expression or equation.

17. The sum of 2 consecutive odd integers is 10.
18. The product of two consecutive even integers is 48.
19. Thirty-five percent of a number.
20. One side of a rectangle is 3 feet longer than the other and the perimeter is 26 feet.

CHAPTER II - TEST B

I. Solve for x in each of the following equations.

1. $bx - c = 0$
2. $b^2 - 4acx = 0$
3. $\frac{1}{4}x + \frac{1}{4}y = 3$
4. $(ax - ay) = 10$
5. $0.02x + 2(0.01x - y) = 3y$

II. Solve each inequality

6. $4 - 8x < 12$
7. $x - 12 > 1$
8. $-(4/17)x \leq 4$
9. $5 - 2(x - 3) < -5x + 5$

10. $-1 < 3 - 2x < 1$ 11. $3 < (4 - 3x)/3 < 7$

12. $0.3 < (0.2x + 4)/2 < 0.5$ 13. $\frac{1}{2}x - 1 < (1/3)x + 1 < \frac{1}{4}x + 2$

14. $0.06x + 0.03 < 0.36x + 0.05$ 15. $4x + 9 \geq 9(x - 3)$

16. $-(1/5)x \leq 30$

III. For each graph write an inequality that has the solution shown by the graph.

17.

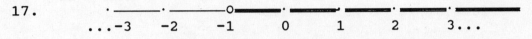

    ...-3   -2   -1   0   1   2   3...

18.

     ... -10   10...

19.

     ...-2   -3   -4   -5   -4   -3   -2 ...

20.

     ... -5   0   5 ...

CHAPTER II - TEST C

I. Solve each inequality.

1. $-5 < 6 - 4x < 5$ 2. $-0.2 \leq 2(-1 - x) \leq 0.2$

3. $\frac{1}{4} \leq 11/(11-x)$, $x \neq 11$ 4. $3x - 5 > 4$

5. $12 - x < 7 - 2x$

II. Solve each word problem.

6. The total cost (including tax) of a farm is $13,440. If the sales tax rate is 12%, how much do you pay without tax ?

7. The sum of two integers is 32. One of the integers is three times the other. Find the integers.

8. The sum of the three angles in a triangle is 180^{0}. The second angle is 10 degrees less than the first and the third angle is three times the first. Find each angle in the triangle.

9. A man left one-third of his property to his wife, one-half to his son, and the remainder, which was $1,500, to his uncle. How much money did the man leave ?

10. A textbook and a notebook costs $20.20. If the textbook costs $18.00 more than the notebook, how much did each cost ?

CHAPTER II - TEST A - SOLUTIONS

I. Solve each equation

1. $x = 2$ 2. $x = 2$ 3. $x = 32$ 4. $x = 20$

5. $x = -2$ 6. $x = -(25/9)$ 7. $x = 8$ 8. $x = 2$

9. $x \neq 3$ 10. $x = 1$ 11. $x = 1000$ 12. $x = -160$

13. inconsistent   14. identity   15. $x = 0$ 16. identity

II. Encode each of the following phrases or sentences as an algebraic expression or equation.

17. x + (x + 2) = 10   18. x·(x + 2) = 48   19. 0.35·x
20. 2[x + (x + 3)] = 26

CHAPTER II - TEST B - SOLUTIONS

I. Solve for x in each of the following equations.

1. x= c/b   2. x = $(b^2)/(4ac)$ 3. x = 12 - y
4. x = (10 + ay)/a   5. x = 125y

II. Solve each inequality

6. x > -1   7. x > 13   8. x ≥ -17
9. x < -2   10. 1 < x < 2   11. -(17/3) < x < -(5/3)
12. -17 < x < -15   13. x < 12   14. x > -(1/15)
15. x ≤ 7.2   16. x ≥ -150   17. x > -1
18. -10 < x ≤ 10   19. x < -5   20. -5 ≤ x < 5

CHAPTER II - TEST C - SOLUTIONS

I. Solve each inequality.

1. ¼ < x < 11/4   2. -1.1 ≤ x ≤ -0.9
3. x ≥ -33   4. x > 3   5. x < -5

II. Solve each word problem.

6. x + 0.12x = 13440   7. Let x be one integer
   1.12x = 13440   x + 3x = 32
   x = $12,000   4x = 32
                                x = 8
                                Two integers are 8, 24

8. A + (A - 10) + 3A = 180
   5A = 190
   A = 38 degrees, A-10 = 28 degrees, and 3A = 114 degrees
   The angles are 38 degrees, 28 degrees, and 114 degrees

9. Let x be the money he left
   x - [(1/3)·x + ½·x] = 1500
   x - 5/6x = 1500
   x = $9,000

10. Let x be the cost of the notebook
    x + (x + 18) = 20.2
    2x = 2.2
    x = $1.1, $1.1 + $18.0 = $19.1
    Text costs $19.1 and the notebook costs $1.1.

◆◆◆◆◆

CHAPTER 3

EXPONENTS AND POLYNOMIALS

3.1 ADDITION and SUBTRACTION of POLYNOMIALS

OBJECTIVES:

*(A) To add two polynomials
*(B) To subtract two polynomials

** REVIEW

A polynomial is defined as an algebraic expression, which may be written as a single term or a finite sum of terms, that contains only nonnegative integer exponents.

Examples:
$3x^5 - 4x^2y^2 + \frac{1}{2}x - y$ and $5x^3 - 2x + 5$ are polynomials.
$3x^{-2} + 2$, $x\sqrt{x}$, and $-5x^{\frac{1}{2}} + 19$ are not polynomials.

Terminology:
• A polynomial with __one__ term is called a __monomial__.
• A polynomial with exactly __two__ terms is called a __binomial__.
• A polynomial with exactly __three__ terms is called a __trinomial__.
• The __degree__ of a __polynomial__ in one variable is the __highest power__ of the variable in the polynomial.

Examples:
    a. The polynomial $5x^4 - 2x$ is a binomial whose degree is 4.
    b. The polynomial $-3x$ is a monomial whose degree is 1.
    c. The polynomial 25 is a monomial whose degree is 0.
    d. The polynomial $-38x^6 - 2x + 5$ is a trinomial whose degree is 6.
    e. The polynomial 0 is a polynomial __without__ a degree.

*(A) To add two polynomials, add the like terms.

Examples:
    a. To add the polynomials

$$6x^3 - x + 5 \quad \text{and} \quad 3x^5 - 4x^3 + 14x - 10,$$

Combine __like terms__ by using the commutative and associative properties.

$$(6x^3 - x + 5) + (3x^5 - 4x^3 + 14x - 10)$$

$$= 3x^5 + (6x^3 - 4x^3) + (-x + 14x) + (5 - 10)$$

$$= 3x^5 + 2x^3 + 13x - 5$$

We may also write the addition vertically.

x^5term	x^4term	x^3term	x^2term	x^1term	constant-term
		$6x^3$		$-x$	$+ 5$
$3x^5$		$-4x^3$		$14x$	-10
$3x^5$		$+2x^3$		$+13x$	-5

b. $(-x^4 + x - 3) + (-3x^4 - 3x^3 + 9x + 20)$

Combine like terms and arrange the exponents in descending order.

$(-x^4 - 3x^4) - 3x^3 + (x + 9x) + (-3 + 20)$

$= -4x^4 - 3x^3 + 10x + 17$

x^4term	x^3term	x^2term	xterm	constant-term
$-1x^4$			$+x$	-3
$-3x^4$	$-3x^3$		$+9x$	$+20$
$-4x^4$	$-3x^3$		$+10x$	$+17$

*(B) Subtraction of two polynomials is accomplished by changing the sign of every term in the second polynomial and adding two polynomials.

Example:
Subtract $(x^3 - 12x^2 + 5x - 19)$ from $(-4x^2 - 7x - 10)$.

$(-4x^2 - 7x - 10) - (x^3 - 12x^2 + 5x + 19)$

$= (- 4x^2 - 7x - 10) - x^3 + 12x^2 - 5x - 19$
Change the sign of every term in the second polynomial

$= -x^3 + (-4x^2 + 12x^2) + (-7x - 5x) + (-10 - 19)$

$= -x^3 + 8x^2 - 12x - 29$

If we write this subtraction vertically.

	x^3term	x^2term	xterm	constant-term		
		$-4x^2$	$-7x$	-10	-->	$- 4x^2 - 7x - 10$
-)	x^3	$-12x^2$	$+5x$	$+19$	-->	$- x^3 + 12x^2 - 5x - 19$
	$-x^3$	$+8x^2$	$-12x$	-29		

Change the sign of every term in the second polynomial and perform addition.

◆◆◆◆◆

38

** EXERCISES

I. True or false ?

1. In the polynomial $-3x^4 + 5x^2 - 10$, the coefficient of x^4 is -3.
2. In the polynomial $-3^4 - x^3 - x^2 - x + 22$,
   the coefficient of x^5 is 1.
3. The degree of the polynomial $-2x^3 + 5x^2 - 5 + x^7$ is 3.
4. The polynomial $3x^5$ is a monomial.
5. The polynomial $x^3 - 2$ is a trinomial.

II. Determine the coefficient of x in each polynomial.

6. $3 - x^3 + 5x$ 7. $3x^5 - 10x^4 + x^3 + 5$
8. 3 9. $\frac{1}{4}x^3 - \frac{1}{2}x - 5$ 10. $9x - 3x^2$

III. Identify each polynomial as monomial, binomial, or trinomial. Give the degree of each.

11. $-5x^7$ 12. 0 13. $p^3 + p^2 + p$ 14. $b^2 - b$

IV. Perform the indicated operations.

15. $(3x^3 - 2x + 1) + (-2x^2 - 5x + 1)$

16. $(a^2 + 5a) + (a^3 - a^2 - a - 1)$

17. $(2.5x^2 - 0.12x + 1) - (-1.5x^2 + 0.88x - 1)$

18. $(x^2 - x) - (x^2 - 2x)$

19. $(x^5 + 4x^4 - 3x^3 - 2x^2 - 1) - (-2x^5 - 3x^4 - 2x^3 - x^2 - 1)$

20. $(\frac{1}{2}x^3 - \frac{1}{4}x^2 - x + 1) - (\frac{1}{4}x^3 - \frac{1}{2}x^2 - \frac{1}{4}x - \frac{1}{4})$

◆◆◆◆◆

** SOLUTIONS TO EXERCISES

1. True.
2. False. (The coefficient of x^5 is 0)
3. False (The degree is 7)   4. True.
5. False. (binomial)   6. 5
7. 0   8. 0
9. $-\frac{1}{2}$ 10. 9
11. monomial, the degree is 7   12. monomial, no degree
13. trinomial, the degree is 3   14. binomial, the degree is 2
15. $3x^3 - 2x^2 - 7x + 2$ 16. $a^3 + 4a - 1$
17. $4x^2 - x + 2$ 18. x
19. $3x^5 + 7x^4 - x^3 - x^2$ 20. $\frac{1}{4}x^3 + \frac{1}{4}x^2 - \frac{3}{4}x + 5/4$

◆◆◆◆◆

3.2 MULTIPLICATION of POLYNOMIALS

OBJECTIVES:

*(A) To multiply two monomials.
*(B) To multiply a monomial and a polynomial.
*(C) To multiply two polynomials.

** REVIEW

*(A) To multiply two monomials use the product rule for multiplication and the commutative and/or associative properties.

Examples:

a. $(3x^5)(2x^2)$

$= (3 \cdot 2) \cdot (x^5 \cdot x^2)$ -->   commutative property

$= 6x^5 x^2$ -->   simplify

$= 6x^7$ -->   product rule

b. $(-3x^2y^3)(5xy^2)$

$= (-3 \cdot 5)(x^2y^3 \cdot xy^2)$ -->   commutative property

$= -15(x^2 \cdot x \cdot y^3 \cdot y^2)$ -->   commutative property

$= -15(x^3y^5)$ -->   product rule

*(B) To multiply a monomial and a polynomial, use the distributive property.

Examples:

a. $(-3x^2)(5x^3 - 2x + 1)$

$= (-3x^2)(5x^3) - (-3x^2)(2x) + (-3x^2)(1)$ distributive property

$= -15x^5 + 6x^3 - 3x^2$ simplify

b. $(p - q - r)(-p)$

$= p \cdot (-p) - q \cdot (-p) - r \cdot (-p)$

$= -p^2 + qp + rp$

*(C) To multiply polynomials, multiply the first polynomial by every term of the second polynomial, then combine like terms.

Examples:

a. $(x^2 + 1)(x + 5)$

$= (x^2 + 1) \cdot x + (x^2 + 1) \cdot 5$ -->   distributive property

$= (x^2 \cdot x) + (1 \cdot x) + (x^2 \cdot 5) + (1) \cdot (5)$ --> distributive property

$= x^3 + x + 5x^2 + 5$ --> simplify and combine like terms

$= x^3 + 5x^2 + x + 5$ --> arrange the exponents in descending order

• The opposite of a polynomial is obtained by multiplying the polynomial by -1.

Examples:

The opposite of $(x^2 - x + 1)$ is $(-1) \cdot (x^2 - x + 1) = -x^2 + x - 1$

The opposite of $(-a^4 - 1)$ is $(-1) \cdot (-a^4 - 1) = a^4 + 1$

♦♦♦♦♦

** EXERCISES

I. True or False ?

1. $-3x^2 \cdot x = -3x^2$ for any value of x

2. $3x^0 \cdot x^7 = 3x^7$ for any value of x

3. $(-3x^3)(9x^2 - 5x - 1) = -27x^6 + 15x^3 + 3x^3$ for any value of x

4. $-(-3x + 5) = 3x - 5$ for any value of x

5. $-(-(4a^2 - 4a + 1)) = 4a^2 - 4a + 1$

II. Find the products

6. $(-2x)^3$ 7. $(2y^2)^3$ 8. $a^7 \cdot b^5 5a^2$

9. $(-9a) \cdot (-7a^2)(-6a^5b^2)$ 10. $(3x^2y^3z) \cdot (-2xyz)$

III. Find the products

11. $(-2x)(x^5 - x - 1)$ 12. $(y^3 - y^2 - y - 1)(-3y^3)$

13. $(x+1)(x^2 - 9x)$ 14. $(-3y^2)(y^5 - y^4 + 5y^2)$

15. $(t - 2)(t + 2)$ 16. $(a-b)(a-b)$

IV. Multiply the following polynomials

17. $\begin{array}{r} x^3 - 2x^2 + x + 5 \\ 3x^2 \\ \hline \end{array}$ 18. $\begin{array}{r} -a - b \\ a + b \\ \hline \end{array}$

19. $\begin{array}{r} -3x^3 - x + 1 \\ x^5 - 4x^2 \\ \hline \end{array}$ 20. $\begin{array}{r} a - b \\ a^2 + ab + b^2 \\ \hline \end{array}$

♦♦♦♦♦

1. False. $(-3x^2 \cdot x = -3x^3)$ 2. True.

3. False. $(-27x^5 + 15x^4 + 3x^3)$ 4. True.   5. True.

6. $-8x^3$ 7. $8y^6$ 8. $5a^9b^5$ 9. $-378a^8b^2$

10. $-6x^3y^4z^2$ 11. $-2x^6 + 2x^2 + 2x$

12. $-3y^6 + 3y^5 + 3y^4 + 3y^3$ 13. $x^3 - 8x^2 - 9x$

14. $-3y^7 + 3y^6 - 15y^4$ 15. $t^2 - 4$

16. $a^2 - 2ab + b^2$ 17. $3x^5 - 6x^4 + 3x^3 + 15x^2$

18. $-(a^2 + 2ab + b^2)$ 19. $-3x^8 - x^6 + 13x^5 + 4x^3 - 4x^2$

20. $a^3 - b^3$

◆◆◆◆◆

3.3 <u>MULTIPLICATION of BINOMIALS</u>

<u>OBJECTIVES</u>:

*(A) To multiply two binomials by using the **FOIL** method.

** <u>REVIEW</u>

The letter F, O, I, and L denote the products of the <u>F</u>irst, <u>O</u>uter, <u>I</u>nner, and <u>L</u>ast terms, respectively.

Example: a.

First _____ Last

$(x + 5)(x + 2)$

Inner

Outer

$(x + 5)(x + 2) = $ **F** $+$ **O** $+$ **I** $+$ **L**

$= x \cdot x + x \cdot 2 + 5 \cdot x + 5 \cdot 2$

$= x^2 + 7x + 10$

b. $(x^2 - 5)(x^3 + 6) = $ First $+$ Outer $+$ Inner $+$ Last

$= x^2 \cdot x^3 + x^2 \cdot 6 + (-5) \cdot x^3 + (-5) \cdot 6$

$= x^5 - 5x^3 + 6x^2 - 30$

◆◆◆◆◆

** EXERCISES

I. True or False ?

   1. $(x - 3)(x + 1) = x^2 - 3$

   2. $(a - 1)(b - 1) = a + b + 1$

   3. $(3a - 2)(-a + 3) = -3a^2 + 11a - 6$

   4. $(3x^3 - 1)(3x^3 + 1) = 9x^6 - 1$

   5. $(x + 4)(y - 4) = xy - 16$

   6. $(x - 2)(2 - x) = x^2 - 4$

   7. $(3x^5 - 1)(2x^7 + 1) = 6x^{12} - 2x^7 + 3x^5 - 1$

   8. $(3a^3 + a)(a - 3a^3) = -9a^6 + a^2$

   9. $(y - 10)(-y + 10) = -y^2 + 20y^2 - 100$

   10. $(x + a)(x - b) = x^2 + ax - bx - ab$

II. Use FOIL method to find each product.

   11. $(x^2 - 6)(x^2 + 2)$ 12. $(-2x^5 + x)(5x^5 - x)$

   13. $(y^5 - 1)(y^5 + 1)$ 14. $(y^3 - 4)(y - 4)$

   15. $(2x^2 + 7)(2x^2 + 7)$ 16. $(3x - 1)(3x^2 - 1)$

   17. $(3x + 4y)(4x + 3y)$ 18. $(h - 5)(5 - h)$

   19. $(10 - x)(10x - 0x^2)$ 20. $(3x^5 - 1)(6x^9 + 1)$

♦♦♦♦♦

** SOLUTIONS TO EXERCISES

1. False. $(x-3)(x+1)=x^2-2x-3$ 2. False. $(a-1)(b-1)=ab-a-b+1$

3. True.   4. True.   5. False. $(x+4)(y-4)=xy-4x+4y-16$

6. False. $(x-2)(2-x)=-[(x-2)(x-2)]=-x^2+4x-4$

7. True.   8. True.   9. True.   10. True.

11. $(x^2 - 6)(x^2 + 2)$ 12. $(-2x^5 + x)(5x^5 - x)$

   $= x^4 - 4x^2 - 12$ $= -10x^{10} + 7x^6 - x^2$

13. $(y^5 - 1)(y^5 + 1)$

$= y^{10} - 1$

14. $(y^3 - 4)(y - 4)$

$= y^4 - 4y^3 - 4y + 16$

15. $(2x^2 + 7)(2x^2 + 7)$

$= 4x^4 + 28x^2 + 49$

16. $(3x - 1) \cdot (3x^2 - 1)$

$= 9x^3 - 3x^2 - 3x + 1$

17. $(3x + 4y)(4x + 3y)$

$= 12x^2 + 25xy + 12y^2$

18. $(h - 5)(5 - h)$

$= -h^2 + 10h - 25$

19. $(10 - x)(10x - 0x^2)$

$= -10x^2 + 100x$

20. $(3x^5 - 1)(6x^9 + 1)$

$= 18x^{14} - 6x^9 + 3x^5 - 1$

◆◆◆◆◆

3.4   SPECIAL PRODUCTS

OBJECTIVES:

*(A) To solve the square of a binomial.
*(B) To compute the product of a sum and a difference.

** REVIEW

*(A) The square of a binomial is solved by using the method illustrated below.

Examples:

a. $(x + y)^2 = (x + y)(x + y) = x^2 + xy + xy + y^2 = x^2 + 2xy + y^2$

square of the **first** term —————————————↑
add twice the **product** of the **two terms**——↑
square of the **last** term ——————————————↑

b.
$$(x - y)^2 = x^2 - 2xy + y^2$$

square of **the first** term ——————————↑
subtract twice the **product** of the **two** terms ——↑
square of the **last term** ———————————↑

c. $(3x + 2y)^2 =$ **square** of the first term $(3x)^2$
$+$ twice the **product** of the **two** terms $(2 \cdot (3x)(2y))$
$+$ **square** of last term $(2y)^2$
$= 9x^2 + 12xy + 4y^2$

d. $(3x - 2y)^2 = (3x)^2 - 2(3x \cdot 2y) + (2y)^2$
$= 9x^2 - 12xy + 4y^2$

44

*(B) The **product** of a **sum** and a **difference** of the **same two terms** is **equal to the difference of two squares.**

Examples:

a. $(a + b)(a - b) = a^2 - b^2$

square of first term ⎯
subtract ⎯
square of second term ⎯

b. $(3x - 5)(3x + 5)$
= square of the first term - square of the second term
= $(3x)^2 - (5)^2$
= $9x^2 - 25$

♦♦♦♦♦

** EXERCISES

I. True or false ?

1. $(-x + 1)^2 = (-x)^2 + 1$ for any value of x

2. $(x - 5)^2 = x^2 - 10x + 25$ for any value of x

3. $(x + 2)(x - 2) = x^2 + 4$ for any value of x

4. $(-3 + 5)^2 = (-3)^2 + 2(-3)(5) + (5)^2$

5. $(9 - 4)(9 + 4) = (9)^2 - (4)^2$

6. $39 \cdot 41 = (40 - 1)(40 + 1) = (40)^2 - (1)^2 = 1600 - 1 = 1599$

7. $(2x + 9)^2 = 4x^2 - 36x + 81$

II. Find the product.

8. $(2p - 3)^2$ 9. $(-r - 10)(-r + 10)$

10. $(5a - b)(a - 5b)$ 11. $(5x - 1)^2$ 12. $(6a - 1)^2$

13. $(y^4 - 1)(y^4 + 1)$ 14. $(-a - b)^2$

15. $(5x + 7y)(7x - 5y)$ 16. $(8x + 1)^2$ 17. $(2.5x - 1)^2$

18. $(3n + 4)(3n - 4)$ 19. $(2x^4 - x^2)^2$ 20. $(8x - b)(b - 4x)$

♦♦♦♦♦

** SOLUTIONS TO EXERCISES

1. False. $(-x+1)^2 = x^2 - 2x + 1$ 2. True.

3. False. $(x+2)(x-2) = x^2 - 4$ 4. True.

5. True.   6. True.

7. False. $(2x+9)^2 = 4x^2 + 36x + 81$ 8. $(2p - 3) = 4p^2 - 12p + 9$

9. $(-r - 10)(-r + 10) = (-r)^2 - (10)^2 = r^2 - 100$

10. $(5a - b)(a - 5b) = 5a^2 - 26ab + 5b^2$

11. $(5x - 1)^2 = 25x^2 - 10x + 1$

12. $(6a - 1)^2 = 36a^2 - 12a + 1$

13. $(y^4 - 1)(y^4 + 1) = (y^4)^2 - 1 = (y^4) \cdot (y^4) - 1 = y^8 - 1$

14. $(-a - b)^2 = ((-a) - (b))^2 = (-a)^2 - 2(-a)(+b) + (-b)^2$
    $= a^2 + 2ab + b^2$

15. $(5x + 7y)(7x - 5y) = 35x^2 + 24xy - 35y^2$

16. $(8x + 1)^2 = 64x^2 + 16x + 1$

17. $(2.5x - 1)^2 = (2.5x)^2 - 2(2.5)x + 1 = 6.25x^2 - 5x + 1$

18. $(3n + 4)(3n - 4) = 9n^2 - 16$

19. $(2x^4 - x^2)^2 = (2x^4)^2 - 2(2x^4)(x)^2 + (x^2)^2 = 4x^8 - 4x^6 + x^4$

20. $(8x - b)(b - 4x) = -32x^2 + 12xb - b^2$

♦♦♦♦♦

3.5 DIVISION of POLYNOMIALS

OBJECTIVES:

* (A) To divide polynomials
* (B) To divide a polynomial by a monomial
* (C) To divide a polynomial by a binomial

** REVIEW

* (A) Rule for dividing exponential expressions:
     If m and n are integers and a≠0, then

$$\frac{a^m}{a^n} = a^{m-n}, \quad Note: a^0=1, \quad a^{-k}=\frac{1}{a^k}$$

Examples:

a. $\dfrac{y^{12}}{y^2} = y^{12-2} = y^{10}$

b. $\dfrac{y^3}{y^{10}} = y^{3-10} = y^{-7} = \dfrac{1}{Y^7}$

c. $\dfrac{2a^3b^4}{6a^2b^5} = \dfrac{1}{3}a^{3-2}b^{4-5} = \dfrac{1}{3}ab^{-1} = \dfrac{a}{3b}$

d. $(-12x^2y^5) \div (-6x^2y^2) = \dfrac{-12x^2y^5}{-6x^2y^2} = 2y^3$

*(B) Rule for dividing a polynomial by a monomial:
     Divide **each term** of the polynomial by the monomial

Examples:

a. $(5x^2 + 19x - 4) \div (x + 4) = ?$

```
              5x  -  1            --> quotient
            _____
   x + 4  ) 5x² + 19x - 4         --> dividend (arrange the terms in
             5x² + 20x                descending order of the exponents)
            _____
                  - x - 4
                  - x - 4
                 _____
                         0       --> remainder
```

47

b. $(-3x^2 + 19x - 9) \div (x + 3) = ?$

$$
\begin{array}{r}
-3x \quad + 28 \qquad \longrightarrow \text{quotient} \\
x{+}3 \enclose{longdiv}{\;-3x^2 + 19x \;\; - 9} \quad \longrightarrow \text{dividend} \\
-3x^2 - \;\; 9x \qquad\qquad\quad \\
\hline
28x - 9 \qquad\quad \\
28x +84 \qquad\quad \\
\hline
-93 \quad \longrightarrow \text{remainder}
\end{array}
$$

$-3x^2+19x-9 = (-3x+28)(x+3)+(-93)$

$\dfrac{-3x^2+19x-9}{x+3} = (-3x+28)+\dfrac{(-93)}{x+3}$

$\dfrac{Dividend}{Divisor} = Quotient + \dfrac{Remainder}{Divisor}$

$Dividend = (Quotient \cdot Divisor) + Remainder$

◆◆◆◆◆

** <u>EXERCISES</u>

I. True or False ?

1. $\dfrac{10^5}{10^2}=10^3$

2. $\dfrac{10x+5}{5}=2x$

3. $\dfrac{3x+1}{3x+1}=1$ *for any value of x*

4. $\dfrac{3x^4}{2x^6}=\dfrac{1}{x^2}$

5. $\dfrac{-a^7b^2c^6}{a^6b^2c^9}=-ac^{-3}$

6. $\dfrac{(-3ab^3)}{(-12a^3b)}=\dfrac{b^2}{4a^2}$

II. Find the quotients.

7. $\dfrac{14x^2-6x}{x}$

8. $\dfrac{15b-40}{5}$

9. $\dfrac{17ab^5}{102a^5b^5}$

10. $\dfrac{-12x^2y^2+3xy^2-36x^2y+9xy}{3xy}$

11. $(3x^4 + 20x^2 + 1) \div (x^2 - 1)$

48

12. $(3x^3 + 6x^2 + 5x - 9) \div (x - 3)$

13. $(x^4 - x^2 - 1) \div (x - 1)$

14. $(8x^3 - 64) \div (2x - 4)$

15. $(a^3 - 512) \div (a - 2)$

III. Write each expression in the form:

$$\text{quotient} + \frac{\text{remainder}}{\text{divisor}}$$

16. $\dfrac{a^2-5}{a}$ 17. $\dfrac{8x^3-1}{x+2}$

18. $\dfrac{x^5}{x-1}$ 19. $\dfrac{9x^2-x+1}{x-2}$

20. $\dfrac{2a-8}{a}$

♦♦♦♦♦

* **<u>SOLUTIONS TO EXERCISES</u>**

1. True.

2. False. $(10x/5) + (5/5) = 2x + 1$

3. False. $x \neq 1/3$

4. False. $(3/2)x^{-2}$

5. True.

6. True.

7. $14x - 6$

8. $3b - 8$

9. $(1/6)a^{-4}$

10. $-4xy + y - 12x + 3$

11.

$$
\begin{array}{r}
3x^2 + 23 \quad \longrightarrow \text{ quotient} \\
x^2 - 1 \overline{)\ 3x^4 + 20x^2 + 1} \quad \longleftarrow \\
3x^4 - 3x^2 \quad \longleftarrow \text{ subtract this term from \underline{the term above it}} \\
\overline{\ 23x^2 + 1} \quad \longleftarrow \\
23x^2 - 23 \quad \longleftarrow \text{ subtract this term from \underline{the term above it}} \\
\overline{\ 24} \quad \longrightarrow \text{ remainder}
\end{array}
$$

49

12.

$$
\begin{array}{r}
3x^2 \;\; + 15x \;\; + 50 \qquad \text{--> quotient} \\
x - 3 \,\overline{\big)\; 3x^3 \;\; + 6x^2 + 5x - 9} \\
3x^3 \;\; - 9x^2 \qquad\qquad\qquad \\
\hline
15x^2 + \;\; 5x - 9 \\
15x^2 - 45x \qquad \\
\hline
50x - 9 \\
50x - 150 \\
\hline
141 \qquad \text{--> remainder}
\end{array}
$$

13.

$$
\begin{array}{r}
x^3 + x^2 \qquad\qquad \text{--> quotient} \\
x - 1 \,\overline{\big)\; x^4 + 0x^3 - x^2 - 1} \\
x^4 - x^3 \qquad\qquad\qquad \\
\hline
x^3 - x^2 - 1 \\
x^3 - x^2 \qquad \\
\hline
-1 \qquad \text{--> remainder}
\end{array}
$$

14.

$$
\begin{array}{r}
4x^2 + 8x + 16 \qquad \text{--> quotient} \\
2x - 4 \,\overline{\big)\; 8x^3 \qquad\qquad - 64} \\
8x^3 - 16x^2 \qquad\qquad \\
\hline
16x^2 \qquad - 64 \\
16x^2 - 32x \qquad \\
\hline
32x - 64 \\
32x - 64 \\
\hline
0 \qquad \text{--> remainder}
\end{array}
$$

15.

$$
\begin{array}{r}
a^2 \;\; + 2a + 4 \qquad\qquad \text{--> quotient} \\
a - 2 \,\overline{\big)\; a^3 \qquad\qquad\qquad - 512} \\
a^3 - 2a^2 \qquad\qquad\qquad \\
\hline
2a^2 \qquad\qquad - 512 \\
2a^2 - 4a \qquad\qquad \\
\hline
4a - 512 \\
4a - \;\;\; 8 \\
\hline
-504 \qquad \text{--> remainder}
\end{array}
$$

50

16. $\dfrac{a^2-5}{a} = \dfrac{a^2}{a} - \dfrac{5}{a} = a - \dfrac{5}{a}$

17. $\dfrac{8x^3-1}{x+2} = 8x^2 - 16x + 32 - \dfrac{65}{x+2}$

18. $\dfrac{x^5}{x-1} = (x^4 + x^3 + x^2 + x + 1) + \dfrac{1}{x-1}$

19. $\dfrac{9x^2-x+1}{x-2} = (9x+17) + \dfrac{35}{x-2}$

20. $\dfrac{2a-8}{a} = \dfrac{2a}{a} - \dfrac{8}{a} = 2 - \dfrac{8}{a}$

♦♦♦♦♦

CHAPTER III - TEST A

Perform the indicated operations.

1. $(3x - 4) - (4 - 3x)$

2. $(2a + 5) + (7a - 5)$

3. $(x^2 - 5x + 1) + (x^2 - 1)$

4. $(-2x^2 + 4) - (4x^2 - 2)$

5. $(x^3 + x^2 + 1) + (x^2 - 1)$

6. $(-10a^4)^3$

7. $(-3a^3t^2) \cdot (- 5 a^2t^3)$

8. $8x - 3x(x^2 + 2x - 1)$

9. $5 - (5 - x - x^2)$

10. $(9x^2 + x) - (3x^3 - 4x^2 + 1)$

11. $(2y + 1)(2y + 1)$

12. $(x + 3)(x - 3)$

13. $(x^2 - 5)(3x^2 + 7)$

14. $(4 - t)(t - 4)$

15. $(2t + 3)(3t + 4)$

16. $(x^2 - 9)(x^2 + 9)$

17. $(11x^2 + 11)(12x^2 + 12)$

18. $(4y - 3)(3y - 4)$

19. $(x^2 - 3x + 1) \cdot x^3$

* 20. $(t - 4)(t - 4)(t - 4)$

* Hint: First multiply the first **two** terms and then multiply the result with the third term.

♦♦♦♦♦

51

CHAPTER III - TEST B

I. Find the quotient and remainder

1. $(35x^3) \div (7x^2)$

2. $(-7x^2y^3) \div (-7xy^2)$

3. $(3a^9b^7) \div (-a^4b^2)$

4. $(8 - 7y) \div (-2)$

5. $(3x^4 + x - 1) \div (-4x)$

6. $(x^7 - 3) \div (x^3 + 1)$

7. $(7x^3 - 1) \div (x + 7)$

8. $(3x^2 - 4y^2) \div (4y^2 - 3x^2)$

9. $(a^3 - b^3) \div (a - b)$

10. $(x^5 - 32) \div (x - 2)$

II. Express each of the following in the form of

$$quotient + \frac{remainder}{divisor}$$

11. $\dfrac{x^3-3}{x+1}$

12. $\dfrac{x^3}{x-1}$

13. $\dfrac{-5x}{5-x}$

14. $\dfrac{x^3}{x+2}$

15. $\dfrac{-3x}{x+1}$

16. $\dfrac{3x+4}{x}$

17. $\dfrac{x^3+x^2+x+1}{x-1}$

18. $\dfrac{-4x^2}{x^2+1}$

19. $\dfrac{2-x^2}{x-2}$

20. $\dfrac{x^3-x^2-x-1}{x-1}$

◆◆◆◆◆

CHAPTER III - TEST C

I. Perform the indicated operations.

1. $(-8x + 2) + (-x^3 - x^2 - x - 1)$ 2. $(x^4 + x + 1) + (3x - 4)$

3. $(9x^2 - 1) + (x^5 + x^2 - 1)$

4. $(2x^3 - 5x^2 + x - 1) + (2x^4 - x^3 - x^2 - x - 1)$

5. $(-11x + 9) + (-9 + 11x - x^2)$ 6. $(-4x - 2) - (-x^4 - x^2 + 1)$

7. $(-x^3 - x^2 + 1) - (3x + 4)$

8. $(1 - 9x^2) - (1 - x^2 - x^5)$ 9. $(x^2 - x + 1) - (x^3 - x^2)$

10. $(-10x + 9) - (2 - 5x - x^2)$ 11. $(-8x)(-2x^3 - x^2 - x - 1)$

12. $(3x - 4)(x^4 + x + 1)$ 13. $(9x^2 - 1)(x^5 + x^2 - 1)$

14. $(x^3 + x^2)(x-1)$ 15. $(a - b)(a - b)$

16. $(2a - b)(a + 2b)$ 17. $(a^2 - b^2)(3 - 2b)$

18. $(5x^2 - x + 1) \div (x^2)$ 19. $(27a^3 - 1) \div (3a - 1)$

20. $(a^3 - a^2 - 1) \div (a + 1)$

◆◆◆◆◆

CHAPTER III - TEST A - SOLUTIONS

Perform the indicated operations.

1. $6x - 8$ 2. $9a$ 3. $2x^2 - 5x$

4. $-6x^2 + 6$ 5. $x^3 + 2x^2$ 6. $-1000a^{12}$

7. $15a^5t^5$ 8. $-3x^3 - 6x^2 + 11x$ 9. $x^2 + x$

10. $-3x^3 + 13x^2 + x - 1$ 11. $4y^2 + 4y + 1$

12. $x^2 - 9$ 13. $3x^4 - 8x^2 - 35$ 14. $-t^2 + 8t - 16$

15. $6t^2 + 17t + 12$ 16. $x^4 - 81$

17. $132(x^4 + 2x^2 + 1)$ 18. $12y^2 - 25y + 12$

19. $x^5 - 3x^4 + x^3$ 20. $t^3 - 12t^2 + 48t - 64$

◆◆◆◆◆

CHAPTER III - TEST B - SOLUTIONS

I. Find the quotient and remainder

1. $(35x^3) \div (7x^2) = 5x$

2. $(-7x^2y^3) \div (-7xy^2) = xy$

3. $(3a^9b^7) \div (-a^4b^2) = -3a^5b^5$

4. $(8 - 7y) \div (-2) = -4 + (7/2)y$

5. $(3x^4 + x - 1) \div (-4x) = -(3/4)x^3 - (1/4) + 1/(4x)$

6.

$$
\begin{array}{r}
x^4 - x \\
x^3+1 \overline{\smash{\big)}\ x^7 - 3} \\
\underline{x^7 + x^4} \\
-x^4 - 3 \\
\underline{-x^4 - x} \\
x - 3
\end{array}
$$

7.

$$
\begin{array}{r}
7x^2 - 49x + 343 \\
x+7 \overline{\smash{\big)}\ 7x^3 - 1} \\
\underline{7x^3 + 49x^2} \\
-49x^2 - 1 \\
\underline{-49x^2 - 343x} \\
343x - 1 \\
\underline{343x + 2401} \\
-2402
\end{array}
$$

8.

$$
\begin{array}{r}
-1 \\
4y^2-3x^2 \overline{\smash{\big)}\ -4y^2 + 3x^2} \\
\underline{-4y^2 + 3x^2} \\
0
\end{array}
$$

9.

$$
\begin{array}{r}
a^2 + ab + b^2 \\
a-b \overline{\smash{\big)}\ a^3 - b^3} \\
\underline{a^3 - a^2b} \\
a^2b - b^3 \\
\underline{a^2b - ab^2} \\
ab^2 - b^3 \\
\underline{ab^2 - b^3} \\
0
\end{array}
$$

10.

$$
\begin{array}{r}
x^4 + 2x^3 + 4x^2 + 8x + 16 \\
x-2 \overline{\smash{\big)}\ x^5 - 32} \\
\underline{x^5 - 2x^4} \\
2x^4 - 32 \\
\underline{2x^4 - 4x^3} \\
4x^3 - 32 \\
\underline{4x^3 - 8x^2} \\
8x^2 - 32 \\
\underline{8x^2 - 16x} \\
16x - 32 \\
\underline{16x - 32} \\
0
\end{array}
$$

11.

$$
\begin{array}{r}
x^2 - x + 1 \\
x+1 \overline{\smash{\big)}\ x^3 - 3} \\
\underline{x^3 + x^2} \\
-x^2 - 3 \\
\underline{-x^2 - x} \\
x - 3 \\
\underline{x + 1} \\
-4
\end{array}
$$

12.

$$
\begin{array}{r}
x^2 + x + 1 \\
x-1 \overline{\smash{\big)}\ x^3} \\
\underline{x^3 - x^2} \\
x^2 \\
\underline{x^2 - x} \\
x \\
\underline{x - 1} \\
1
\end{array}
$$

13.

$$
\begin{array}{r}
5 \\
x-5 \overline{\smash{\big)}\ 5x} \\
\underline{5x - 25} \\
25
\end{array}
$$

14.

$$
\begin{array}{r}
x^2 - 2x + 4 \\
x+2 \overline{\smash{\big)}\ x^3} \\
\underline{x^3 + 2x^2} \\
-2x^2 \\
\underline{-2x^2 - 4x} \\
4x \\
\underline{4x+8} \\
-8
\end{array}
$$

15.

$$
\begin{array}{r}
-3 \\
x+1 \overline{\smash{\big)}\ -3x} \\
\underline{-3x - 3} \\
3
\end{array}
$$

16.

$$
\begin{array}{r}
3 \\
x \overline{\smash{\big)}\ 3x + 4} \\
\underline{3x} \\
4
\end{array}
$$

17.
$$x-1 \overline{)\begin{array}{l} x^2 + 2x + 3 \\ x^3 + x^2 + x + 1 \\ \underline{x^3 - x^2} \\ 2x^2 + x + 1 \\ \underline{2x^2 - 2x} \\ 3x + 1 \\ \underline{3x - 3} \\ 4 \end{array}}$$

18.
$$x^2+1 \overline{)\begin{array}{l} -4 \\ -4x^2 \\ \underline{-4x^2 - 4} \\ 4 \end{array}}$$

19.
$$x-2 \overline{)\begin{array}{l} x + 2 \\ x^2 - 2 \\ \underline{x^2 - 2x} \\ 2x-2 \\ \underline{2x-4} \\ 2 \end{array}}$$

20.
$$x-1 \overline{)\begin{array}{l} x^2 - 1 \\ x^3 - x^2 - x - 1 \\ \underline{x^3 - x^2} \\ -x - 1 \\ \underline{-x + 1} \\ -2 \end{array}}$$

11. $(x^2-x+1) + \dfrac{-4}{x+1}$

12. $(x^2+x+1) + \dfrac{1}{(x-1)}$

13. $\dfrac{-5x}{5-x} = \dfrac{5x}{x-5} = 5 + \dfrac{25}{(x-5)}$

14. $(x^2-2x+4) - \dfrac{8}{x+2}$

15. $(-3) + \dfrac{3}{x+1}$

16. $3 + \dfrac{4}{x}$

17. $(x^2+2x+3) + \dfrac{4}{x-1}$

18. $(-4) + \dfrac{4}{x^2+1}$

19. $\dfrac{2-x^2}{x-2} = -\dfrac{x^2-2}{x-2} = -(x+2) - \dfrac{2}{x-2}$

20. $(x^2-1) + \dfrac{-2}{x-1}$

♦♦♦♦♦

CHAPTER III - TEST C - SOLUTIONS

I. Perform the indicated operations.

1. $(-8x + 2) + (-x^3 - x^2 - x - 1) = -(x^3 + x^2 + 9x - 1)$

2. $(x^4 + x + 1) + (3x - 4) = (x^4 + 4x - 3)$

3. $(9x^2 - 1) + (x^5 + x^2 - 1) = (x^5 + 10x^2 - 2)$

4. $(2x^3 - 5x^2 + x - 1) + (2x^4 - x^3 - x^2 - x -1) = (2x^4 + x^3 -6x^2 - 2)$

5. $(-11x + 9) + (-9 + 11x - x^2) = -x^2$

6. $(-4x - 2) - (-x^4 - x^2 + 1) = x^4 + x^2 - 4x - 3$

7. $(-x^3 - x^2 + 1) - (3x + 4) = -x^3 - x^2 - 3x - 3$

8. $(1 - 9x^2) - (1 - x^2 - x^5) = x^5 - 8x^2$

9. $(x^2 - x + 1) - (x^3 - x^2) = -x^3 + 2x^2 - x + 1$

10. $(-10x + 9) - (2 - 5x - x^2) = x^2 - 5x + 7$

11. $(-8x)(-2x^3 - x^2 - x - 1) = 16x^4 + 8x^3 + 8x^2 + 8x$

12. $(3x - 4)(x^4 + x + 1) = 3x^5 - 4x^4 + 3x^2 - x - 4$

13. $(9x^2 - 1)(x^5 + x^2 - 1) = 9x^7 - x^5 + 9x^4 - 10x^2 + 1$

14. $(x^3 + x^2)(x-1) = x^4 - x^2$

15. $(a - b)(a - b) = a^2 - 2ab + b^2$

16. $(2a - b)(a + 2b) = 2a^2 + 3ab - 2b^2$

17. $(a^2 - b^2)(3 - 2b) = 3a^2 - 2a^2b - 3b^2 + 2b^3$

18. $(5x^2 - x + 1) \div (x^2) = 5 + (-x+1)/x^2$

19. $(27a^3 - 1) \div (3a - 1) = (9a^2 + 3a + 1)$

20. $(a^3 - a^2 - 1) \div (a + 1) = (a^2 - 2a + 2) + [-3/(a+1)]$

18.

$$x^2 \overline{)\begin{array}{l} 5 \\ 5x^2 - x + 1 \\ \underline{5x^2} \\ -x + 1 \end{array}}$$

19.

$$3a-1 \overline{)\begin{array}{l} 9a^2 + 3a + 1 \\ 27a^3 - 1 \\ \underline{27a^3 - 9a^2} \\ 9a^2 - 1 \\ \underline{9a^2 - 3a} \\ 3a - 1 \\ \underline{3a - 1} \\ 0 \end{array}}$$

20.

$$a+1 \overline{)\begin{array}{l} a^2 - 2a + 2 \\ a^3 - a^2 - 1 \\ \underline{a^3 + a^2} \\ -2a^2 - 1 \\ \underline{-2a^2 - 2a} \\ 2a - 1 \\ \underline{2a + 2} \\ -3 \end{array}}$$

◆◆◆◆◆

CHAPTER 4

FACTORING

4.1 FACTORING OUT COMMON FACTORS

OBJECTIVES:

*(A) To perform **prime factorization** of integers.
*(B) To find the GCF (Greatest Common Factor) for monomials.
*(C) To factor the GCF (Greatest Common Factor) for an algebraic
       expression.

** REVIEW

*(A) When finding the **prime factorization** of integers, remember
       that a prime number is a positive integer greater than 1
       that has only **two** factors: 1 and itself.

Examples:
   2, 3, 5, 7, 11, 13, 17, 19, 23, 29, 31, 37, 41, 43, 47...
   are examples of prime numbers.

Examples:

   Find the prime factorization of a. 28   b. 24   c. 120
   a. 28 = 2·14 --> 2 is a prime. (14 is not, factor it further)
        = 2·2·7 --> all the factors are prime.

   b. 15 = 3·5 --> both factors are prime.

   c. 120 = 2·60
          = 2·2·30
          = 2·2·2·15
          = 2·2·2·3·5 --> all the factors are prime.

*(B) To find the GCF for monomials, find the prime factorization
       of each. The GCF is the product of all the common factors.

Examples:
   Find the GCF for the following:
     a. 250, 100   b. 12, 33, 45

• Perform the factorization for the numbers and then find
       the greatest common factor.
     a. 250 = 2·125 = 2·5·25 = (2·5·5)·5
        100 = 2·50 = 2·2·25 = (2·5·5)·2
        GCF = 2·5·5 = 50

57

b. $12 = 2 \cdot 6 = 2 \cdot 2 \cdot 3$
   $33 = 3 \cdot 11$
   $45 = 3 \cdot 15 = 3 \cdot 3 \cdot 5$
   GCF $= 3$

- To find the GCF for monomials with variables:
  Step 1. find the GCF for the coefficients of the monomials.
  Step 2. find the GCF for the variable(s) of the monomials.

  The **GCF** for the variable(s) of the monomials is the **variable(s)** that is **common** to all of the monomials, where the degree on that variable is the **lowest degree** found in all of the monomials.

Examples:
  Find the GCF for the following monomials:

   a. $21x^5$, $36x^2$ b. $9x^7y^3$, $27x^2y^7$, and $51x^2y^3$

Solutions:
   a. The GCF of the coefficients 21 and 36 is 3.
      The GCF of the variable x is x^2 (The lowest power of x is 2)
      The GCF of $21x^5$ and $36x^2$ is $3x^2$

   b. The GCF of the coefficients 9, 27, 51 is 3.
      The GCF of the variable x is x^2 (The lowest power of x is 2)
      The GCF of the variable y is y^3 (The lowest power of y is 3)
      The GCD of $9x^7y^3$, $27x^2y^7$, and $51x^2y^3$ is $3x^2y^3$

*(C) To factor GCF out of an algebraic expression:
   a. Find the GCF for the coefficients of the monomials.
   b. Find the GCF for the variable(s) of the monomials.
   c. Factor the GCF for both coefficients and the variables

Examples:
  Find the GCF for the following:

   a. $21x^4 - 15x^3 + 30x^2$ b. $x^4y^3 + x^3y^4$

Solutions:
   a. $21x^4 - 15x^3 + 30x^2$ --> The GCF of the coefficients is 3

      $= 3 \cdot (7x^4 - 5x^3 + 10x^2)$ --> The lowest power of x in
                                               the expression is 2

      $= 3x^2 \cdot (7x^2 - 5x + 10)$

      GCF $= 3x^2$

   b. $x^4y^3 + x^3y^4$ --> The GCF of the coefficients is 1

      $x^4y^3 + x^3y^4$ --> The lowest power of x in the expression is 3.
                                The lowest power of y in the expression is 3.
      GCF $= x^3y^3$

• We can also factor out a common factor with a negative sign
  in front of the expression.

Examples:
$$5x - 5y = 5(x - y) = -5(-x + y) \quad \text{and}$$
$$-7x^3 + 21x^2 - 42x = -7x(x^2 - 3x + 6)$$

♦♦♦♦♦

** EXERCISES

I. True or False ?

1. 110 is a prime number.

2. The prime factorization of 28 is $2 \cdot 13$.

3. The GCF of the integers 14 and 18 is 2.

4. The GCF of the polynomial $x^7y^2z - x^5y^7z^7$ is $x^7y^7z^7$

5. The GCF of $27a^5b^3 - 3a^4b^2$ is $3a^4b^2$

6. $-5x^5 - 15x^4 - 10x^3 = -5x^3(x^2 + 3x + 2)$

II. Find the prime factorization of each integer.

7. 22
9. 200
11. 35

8. 84
10. 121
12. 1000

III. Find the GCF for each group of integers.

13. 12, 24, 36
15. 1000, 10000

14. 66, 77, 88
16. 40, 60, 80

IV. Find the GCF of each group of monomials.

17. $6y$, $9y^2$

18. $4y^2$, $6y^3$, $12y^4$

19. $39a^2b^3$, $13a^3$, $52a^2b^2$

20. $20xy$, $30x^2y$

V. Factor out the GCF for each expression.

21. $x^2 - 9x$

22. $12x^2y + 32xy^2 - 6xy$

23. $-(x+4)t + (x+4)t^2$

24. $3a^2 + 4a^2b + 5a^2b^2$

25. $(a+b)^3x + (a+b)^3y$

♦♦♦♦♦

** <u>SOLUTIONS TO EXERCISES</u>

I. True / False

1. False. $(110 = 2 \cdot 5 \cdot 11)$
2. False. $(28 = 2^2 \cdot 7)$
3. True.
4. False. $(GCF = x^5 y^2 z)$
5. True.
6. True.

II. Find the prime factorization of each integer.

7. $22 = 2 \cdot 11$
8. $84 = 2^2 \cdot 3 \cdot 7$
9. $200 = 2^3 \cdot 5^2$

10. $121 = 11^2$
11. $35 = 5 \cdot 7$
12. $1000 = 2^3 5^3$

III. Find the GCF of each group of integers.

13. 12, 24, 36
    $12 = 2^2 \cdot 3 = \underline{\textbf{12}} \cdot 1$
    $24 = 2^3 \cdot 3 = \underline{\textbf{12}} \cdot 2$
    $36 = 2^2 \cdot 3^2 = \underline{\textbf{12}} \cdot 3$
    GCF = 12

14. 66, 77, 88
    $66 = 2 \cdot 3 \cdot \underline{\textbf{11}}$
    $77 = 7 \cdot \underline{\textbf{11}}$
    $88 = 2^3 \underline{\textbf{11}}$
    GCF = 11

15. 1000, 10000
    $1000 = \underline{\textbf{1000}} \cdot 1$
    $10000 = \underline{\textbf{1000}} \cdot 10$
    GCF = 1000

16. 40, 60, 80
    $40 = \underline{\textbf{20}} \cdot 2$
    $60 = \underline{\textbf{20}} \cdot 3$
    $80 = \underline{\textbf{20}} \cdot 4$
    GCF = 20

IV. Find the GCF of each group of monomials.

17. $6y, 9y^2$ GCF = 3y

18. $4y^2, 6y^3, 12y^4$ GCF = $2y^2$

19. $39a^2b^3, 13a^3, 52a^2b^2$ GCF = $13a^2$

20. $20xy, 30x^2y$ GCF = 10xy

V. Factor out the GCF of each expression.

21. $x^2 - 9x$ GCF = x

22. $12x^2y + 32xy^2 - 6xy$ GCF = 2xy

23. $-(x+4)t + (x+4)t^2$ GCF = $(x+4)t$

24. $3a^2 + 4a^2b + 5a^2b^2$ GCF = a^2

25. $(a+b)^3x + (a+b)^3y$ GCF = $(a+b)^3$

♦♦♦♦♦

4.2 **FACTORING THE SPECIAL PRODUCTS**

OBJECTIVES:

*(A) To factor the difference of two squares.
*(B) To factor a perfect square trinomial.
*(C) To factor by grouping.

** **REVIEW**

*(A) Consider the **difference of two squares**

$a^2 - b^2 = (a+b)(a-b)$ for any real numbers a and b.

Examples:
  Factor each of the following polynomials.

  a. $9x^2 - 25$ b. $4a^2 - b^2$ c. $100x^2 - y^2$

Solutions:
  a. $9x^2 - 25 = (3x)^2 - (5)^2 = (3x + 5)(3x - 5)$

  b. $4a^2 - b^2 = (2a)^2 - (b)^2 = (2a + b)(2a - b)$

  c. $100x^2 - 169y^2 = (10x)^2 - (13y)^2 = (10x + 13y)(10x - 13y)$

*(B) Consider the **perfect square trinomial**

  For any real numbers a and b

  $(a + b)^2$
   $= a^2$ --> square of the first term
   $+ 2ab$ --> add twice the product of first and the last terms
   $+ b^2$ --> square of the last term

  Therefore $(a+b)^2 = a^2 + 2ab + b^2$ is a perfect square trinomial.
         $(a-b)^2 = a^2 - 2ab + b^2$ is a perfect square trinomial.

Examples:
  Identify each polynomial as a difference of two squares,
  a perfect square trinomial, or neither.

    a. $4a^2 - 4a + 1$ b. $25x^2 + 20x + 16$ c. $49x^2 - 4y^2$

  Solutions:
        a. $4a^2 - 4a + 1 = (2a)^2 - 2(2a) \cdot 1 + (1)^2 = (2a - 1)^2$

Square of the first term ─┘
Twice the product of the ────┘
square roots of the two terms
Square of the last term ──────────┘

    This is the perfect square trinomial.

61

b. $25x^2 + 20x + 16 \neq (5x)^2 + 2(5x) \cdot (4) + (4)^2$
The first term $25x^2$ is the square of $(5x)$.
The last term 16 is the square of (4).
The middle term $20x$ is **not** twice the product of the square
roots of the two terms which should be $2 \cdot (5x) \cdot (4) = 40x$
This is neither a perfect square nor the difference of two squares.

c. $49x^2 - 4y^2 = (7x)^2 - (2y)^2$ --> difference of two squares

*(C) When factoring by grouping remember:

- A prime polynomial is a polynomial which cannot be factored.
- A polynomial is <u>factored completely</u> when it is written as
  a product of prime polynomials.

Examples:
    $(3x - y)$, $(2x^2 + 1)$, and $(x^2 + x + 1)$ are prime polynomials
Examples:
    Factor each polynomial completely.
    Hint: First find the GCF and factor it out

a. $8x^3 - x^2$ b. $12x^3 - 12x^2 + 3x$ c. $-2a^3 + 2ab^2$

Solutions:
    a. $8x^3 - x^2 = x^2(8x - 1)$ --> GCF = x^2

    b. $12x^3 - 12x^2 + 3x = 3x(4x^2 - 4x + 1)$ --> GCF = $3x$
        $= 3x \cdot [(2x)^2 - 2 \cdot (2x) \cdot (1) + (1)^2]$
        This is a perfect square trinomial
        $= 3x(2x - 1)^2$
    c. $-2a^3 + 2ab^2$
        $= -2a(a^2 - b^2)$ --> GCF = $-2a$
        $= -2a(a - b)(a + b)$ --> This is the difference of two squares

- Grouping of terms
  Consider **(ab + bc) + (da + cd)**
  The GCF of (ab + bc) is b and the GCF of (da + cd) is d.
  Therefore (ab + bc) = b(a+ c) and (da + cd) = d(a + c)
  The GCF of b**(a + c)** + d**(a + c)** is (a + c)
  (a**b** + **b**c) + (**d**a + c**d**)   --> factor by grouping
  $= $ **b**$(a + c) + $**d**$(a + c)$
  $= $ **(a + c)**$(b + d)$

Examples:

a. $3x^2 + xz + 2z^2 + 6xz$ b. $6x^3 + 11x^2 - 6x - 11$

Solutions:

a. $3x^2 + xz + 2z^2 + 6xz$
   $= (3x^2 + xz) + (2z^2 + 6xz)$
   $= $ **x**$(3x + z) + $**2z**$(z + 3x)$
   $= $ **(3x + z)**$(x + 2z)$

62

b. $6x^3 + 11x^2 - 6x - 11$
   $= (6x^3 + 11x^2) - (6x + 11)$
   $= x^2(6x + 11) - 1 \cdot (6x + 11) = (6x + 11)(x^2 - 1)$
   $= (6x + 11)(x + 1)(x - 1)$

♦♦♦♦♦

**EXERCISES:

I. Factor each polynomial.

1. $x^2 - 16$ 2. $49x^2 - 4y^2$ 3. $81a^2 - 100b^2$ 4. $144x^2 - 1$

II. Identify each polynomial as a difference of two squares, a perfect square trinomial, or neither of these.

5. $4x^2 + 121y^2$ 6. $a^2 - 9x + 81$

7. $9x^2 - 24x + 16$ 8. $4a^2 - 77$

III. Factor each of the following perfect square trinomials.

9. $4a^2 + 12a + 9$ 10. $121x^2 - 220xy + 100y^2$

11. $9m^2 + 30mn + 25n^2$ 12. $49p^2 + 42pq + 9q^2$

IV. Factor each polynomial completely.

13. $-6x^2 + 9xy$ 14. $6a^4 - 12a^3b + 6a^2b^2$

15. $-3a^2 + 48$ 16. $-5a^3b^2 + 20a^2b^3 - 20ab^4$

V. Use grouping to factor each polynomial completely.

17. $2ac + ad + 4bc + 2bd$ 18. $2xa + 8xb^2 + ay + 4yb^2$

19. $3x^4 - 3x^2y^2 + x^2 - y^2$ 20. $8mp + 10mq + 12tp + 15tq$

♦♦♦♦♦

**SOLUTIONS TO EXERCISES

I. Factor each polynomial.

1. $x^2 - 16 = (x + 4)(x - 4)$

2. $49x^2 - 4y^2 = (7x + 2y)(7x - 2y)$

3. $81a^2 - 100b^2 = (9a + 10b)(9a - 10b)$

4. $144x^2 - 1 = (12x - 1)(12x + 1)$

II. Identify each polynomial as a difference of two squares, a perfect square trinomial, or neither.

5. $4x^2 + 121y^2$ (neither)

6. $a^2 - 9x + 81$ (neither)

7. $9x^2 - 24x + 16 = (3x - 4)^2$ (perfect square trinomial)

8. $4a^2 - 77$ (neither)

III. Factor each of the following perfect square trinomials.

9. $4a^2 + 12a + 9 = (2a + 3)^2$

10. $121x^2 - 220xy + 100y^2 = (11x - 10y)^2$

11. $9m^2 + 30mn + 25n^2 = (3m + 5n)^2$

12. $49p^2 + 42pq + 9q^2 = (7p + 3q)^2$

IV. Factor each polynomial completely.

13. $-6x^2 + 9xy = -3x(2x - 3y)$

14. $6a^4 - 12a^3b + 6a^2b^2 = 6a^2(a^2 - 2ab + b^2) = 6a^2(a - b)^2$

15. $-3a^2 + 48 = -3(a^2 - 16) = -3(a + 4)(a - 4)$

16. $-5a^3b^2 + 20a^2b^3 - 20ab^4 = -5ab^2(a^2 - 4ab + 4b^2)$

       $= -5ab^2(a - 2b)^2$

V. Use grouping to factor each polynomial completely.

17. $2ac + ad + 4bc + 2bd$
   $= a(2c + d) + 2b(2c + d)$
   $= (2c + d)(a + 2b)$

18. $2xa + 8xb^2 + ay + 4yb^2$
   $= 2x(a + 4b^2) + y(a + 4b^2)$
   $= (a + 4b^2)(2x + y)$

19. $3x^4 - 3x^2y^2 + x^2 - y^2$
   $= 3x^2(x^2 - y^2) + 1 \cdot (x^2 - y^2)$
   $= (x^2 - y^2)(3x^2 + 1)$
   $= (x + y)(x - y)(3x^2 + 1)$

20. $8mp + 10mq + 12tp + 15tq$
   $= 2m(4p + 5q) + 3t(4p + 5q)$
   $= (4p + 5q)(2m + 3t)$

◆◆◆◆◆

4.3 FACTORING $ax^2 + bx + c$ with a = 1.

OBJECTIVES:

*(A) To factor trinomials with a leading coefficient of 1.
*(B) To factor completely.

** **REVIEW**

*(A) When factoring trinomials with a leading coefficient of 1,
     consider $x^2 + $ **(a + b)**$x + $ **a·b** = (x + **a**)(x + **b**)

     sum of a and b ⎯⎯⎯⎯⎯
     product of a and b ⎯⎯⎯⎯⎯⎯⎯

Examples:
  Factor the following trinomials

  a. $x^2 + 6x - 16$ b. $x^2 - 13xy - y^2$
  c. $x^2 - 3x - 28$ d. $x^2 - 13xy + y^2$

Solutions:

  a. $x^2 + 6x - 16$
     = (x + a)(x + b)
  where a+b = 6 and a·b = -16.
  a·b = -16   --> you have two possibilities:
  a·b = -16 = **-(2)·(8)** = -(4)·(4),
  but you must have a + b = 6, therefore a = 8, b = -2 is the only choice
  to satisfy the conditions.
  Therefore, $x^2 + 6x - 16 = (x + 8)(x - 2)$

  b. $x^2 - 13xy + 12y^2$
     = (x + ay)(x + by)
  where a+b= -13 and a·b = 12.
  a·b = 12   --> you have three possibilities:
  a·b = 12 = **(-1)·(-12)** = (-2)·(-6) = (-4)·(-3),
  but you must have a + b = -13, therefore a = -1, b = -12
  is the only choice to satisfy the conditions.
  Therefore, $x^2 - 13xy + 12y^2 = (x - y)(x - 12y)$

  c. $x^2 - 3x - 28$
     = (x + a)(x + b)
  where a·b = -28 =(-2)(14) = **(-7)(4)** and a + b = -3 = **(-7)+(4)**
     $x^2 - 3x - 28 = (x + 4)(x - 7)$

  d. $x^2 - 13xy + y^2 = (x + a)(x + b)$
  where a·b = 1 and a + b = -13
  you **can not** find any numbers a and b to satisfy the conditions.
  Therefore, this is a prime trinomial.

*(B) To factor completely, express a polynomial as the products of
     prime polynomials.

Examples:

    a. $x^4 - 8x^3 + 15x^2$ b. $x^3 + 3x^2y + xy^2$

Solutions:

    a. $x^4 - 8x^3 + 15x^2$ --> where $a \cdot b = 15 = 3 \cdot 5 = (-3)(-5)$
                                    --> $a + b = -8 = (-3) + (-5)$
       $= x^2(x^2 - 8x + 15)$ --> therefore, $a = -3$ and $b = -5$

       $= x^2(x + a)(x + b)$

       $x^2(x^2 - 8x + 15) = x^2(x - 3)(x - 5)$

    b. $x^3 + 3x^2y + xy^2$

       $= x(x^2 + 3xy + y^2)$

       where $a \cdot b = 1$ and $a + b = 3$. We cannot find such numbers
       to satisfy the conditions.

       Therefore $(x^2 + 3xy + y^2)$ is a prime polynomial.

       $x^3 + 3x^2y + xy^2 = x(x^2 + 3xy + y^2)$

◆◆◆◆◆

EXERCISES:

I. Factor the trinomials.

 1. $x^2 - 9x + 14$ 2. $x^2 + 2x - 99$ 3. $25 - 10x + x^2$

 4. $a^2 - 5a - 36$ 5. $x^2 + 3x + 2$ 6. $a^2 - 13a + 36$

 7. $a^2 - 5a + 4$ 8. $x^2 - 33xy + 200y^2$ 9. $100 - 25a + a^2$

10. $x^2 - 24xy + 143$

II. Factor each polynomial completely.

11. $x^5 - x^3$ 12. $a^2b + ab^2$

13. $3a^3 - 39a^2 + 108a$ 14. $32w^5 - 2w^3$

15. $9a^4 - 45a^3 - 324a^2$ 16. $x^2y^2 + a^2y^2 + b^2y^2$

17. $-100y^2 + 40xy^2 - 4x^2y^2$ 18. $7t^3 + 21t^2 - 28t$

19. $-6x^2y + 54xy - 84y$ 20. $12w^4 - w^2$

21. $-a^4bc + a^2b^3c$

◆◆◆◆◆

66

**SOLUTIONS TO EXERCISES

I. Factor the trinomials.

1. $x^2 - 9x + 14 = (x - 2)(x - 7)$
    $a \cdot b = \mathbf{14 = (-2)(-7)}$ and $a + b = \mathbf{(-2) + (-7) = -9}$

2. $x^2 + 2x - 99 = (x + 11)(x - 9)$
    $a \cdot b = \mathbf{-99 = (+11)(-9)}$ and $a + b = \mathbf{(+11) + (-9) = 2}$

3. $25 - 10x + x^2 = (x^2 - 10x + 25) = (x - 5)(x - 5) = (x - 5)^2$
    $a \cdot b = \mathbf{25 = (-5)(-5)}$ and $a + b = \mathbf{(-5) + (-5) = -10}$

4. $a^2 - 5a - 36 = (a - 9)(a + 4)$
    $a \cdot b = \mathbf{-36 = (-9)(4)}$ and $a + b = \mathbf{(-9) + (4) = -5}$

5. $x^2 + 3x + 2 = (x + 1)(x + 2)$
    $a \cdot b = \mathbf{2 = (1)(2)}$ and $a + b = \mathbf{3 = (2) + (1) = 3}$

6. $a^2 - 13a + 36 = (a - 9)(a - 4)$
    $a \cdot b = \mathbf{36 = (-9)(-4)}$ and $a + b = \mathbf{-13 = (-9) + (-4)}$

7. $a^2 - 5a + 4 = (a - 4)(a - 1)$
    $a \cdot b = \mathbf{4 = (-4)(-1)}$ and $a + b = \mathbf{-5 = (-4) + (-1)}$

8. $x^2 - 33xy + 200y^2 = (x - 8y)(x - 25y)$
    $a \cdot b = \mathbf{200 = (-8) \cdot (-25)}$ and $a + b = \mathbf{-33 = (-8) + (-25)}$

9. $100 - 25a + a^2 = a^2 - 25a + 100 = (a - 5)(a - 20)$
    $a \cdot b = \mathbf{100 = (-5) \cdot (-20)}$ and $a + b = \mathbf{(-5) + (-20)}$

10. $x^2 - 24x + 143 = (x - 11)(x - 13)$
    $a \cdot b = \mathbf{143 = (-11) \cdot (-13)}$ and $a + b = \mathbf{(-11) + (-13)}$

II. Factor each polynomial completely.

11. $x^5 - x^3 = x^3(x^2 - 1) = x^3(x + 1)(x - 1)$
    Difference of a square

12. $a^2b + ab^2 = ab(a + b)$

13. $3a^3 - 39a^2 + 108a = 3a(a^2 - 13a + 36) = 3a(a - 9)(a - 4)$

14. $32w^5 - 2w^3 = 2w^3(16w^2 - 1) = 2w^3(4w + 1)(4w - 1)$

15. $9a^4 - 45a^3 - 324a^2 = 9a^2(a^2 - 5a - 36) = 9a^2(a - 9)(a + 4)$

16. $x^2y^2 + a^2y^2 + b^2y^2 = y^2(x^2 + a^2 + b^2)$

17. $-100y^2 + 40xy^2 - 4x^2y^2 = -4y^2(25 - 10x + x^2) = -4y^2(x - 5)^2$

18. $7t^3 + 21t^2 - 28t = 7t(t^2 + 3t - 4) = 7t(t + 4)(t - 1)$

19. $-6x^2y + 54xy - 84y = -6y(x^2 - 9x + 14) = -6y(x - 2)(x - 7)$

20. $12w^4 - w^2 = w^2(12w^2 - 1)$

21. $-a^4bc + a^2b^3c = a^2bc(-a^2 + b^2)$

   $= a^2bc(b^2 - a^2) = a^2bc(b - a)(b + a)$

◆◆◆◆◆

4.4 FACTORING $ax^2 + bx + c$ with $a \neq 1$

OBJECTIVES:

*(A) To factor trinomials with a leading coefficient
   that is **not** 1.
*(B) To factor completely.

** REVIEW

*(A) Factor trinomials with a leading coefficient that is **not** 1.

Examples:
  Factor the following trinomials:

  a. $6x^2 + 19x + 15$ b. $10x^2 + 53x - 11$

Solutions:

  a. $6x^2 + 19x + 15$
     $= ax^2 + bx + c = (?x + ?)(?x + ?)$ -->   use the ac method

Step 1.
   Try to find two numbers that have a product equal to ac = (6)(15) and
   the sum equal to b = 19.
   If one number is 9 and the other number is 10,
   then the product = (9)(10) = ac, which is 90, and the sum of the two
   numbers is equal to (9) + (10) = b, which is 19.

Step 2.
   $6x^2 + \mathbf{19x} + 15 = 6x^2 + \mathbf{[+9 + 10]x} + 15$
   facoring by the ac methhod

Step 3.
   $6x^2 + \mathbf{[9 + 10]x} + 10$

   $= (6x^2 + 9x) + (10x + 15)$

   $= 3x(2x + 3) + 5(2x + 3)$ factor by grouping

   $= (2x + 3)(3x + 5)$

68

b. $10x^2 + 53x - 11$

$\quad = ax^2 + bx + c = (?x + ?)(?x + ?)$ --> use ac method

Step 1.
   Try to find two numbers that have a product equal to
   $ac = (10)(-11)$ and   a sum equal to $b = 53$.
   There are three possibilities: $-110 = -(2 \cdot 55) = -(10 \cdot 11) = -(5 \cdot 22)$
   $ac = -110 = (-2) \cdot (55)$ is the only choice to satisfy the sum $b = 53$.

Step 2.
   $10x^2 + 53x - 11$
   $= 10x^2 + \mathbf{[-2 + 55]x} - 11$ --> factor by the ac method

Step 3.
   $10x^2 + \mathbf{[-2 + 55]x} - 11$
   $= (10x^2 - 2x) + (55x - 11)$ --> factor by grouping
   $= 2x(5x - 1) + 11(5x - 1)$
   $= (5x - 1)(2x + 11)$

*(B) Factor completely.

- Factor out the common factor.
- Use the factoring technique shown in *(A) to factor the polynomials completely.

Examples:

a. $2a^5b - 6a^3b^3 - 8ab^5$ b. $24x^3y + 42x^2y^2 - 45xy^3$

Solutions:

a. $2a^5b - 6a^3b^3 - 8ab^5$

   $= 2ab(a^4 - 3a^2b^2 - 4b^4)$ --> common factor **2ab**

   $= 2ab[a^4 + \mathbf{(-4+1)}a^2b^2 - 4b^4]$

   where $a \cdot c = (1) \cdot (-4)$ and $b = a + c = (-4) + 1 = -3$

   $= 2ab[(a^4 - 4a^2b^2) + (a^2b^2 - 4b^4)]$

   $= 2ab[a^2(a^2 - 4b^2) + b^2(a^2 - 4b^2)]$

   $= 2ab(a^2 - 4b^2)(a^2 + b^2)$

   $= 2ab(a + 2b)(a - 2b)(a^2 + b^2)$

b. $24x^3y + 42x^2y^2 - 45xy^3$

$= 3xy(8x^2 + 14xy - 15y^2)$ --> common factor **3xy**
where $a \cdot c = (8) \cdot (-15) = -120 = -(2 \cdot 60) = -(4 \cdot 30) = -(8 \cdot 15)$
$= -(24 \cdot 5) = \mathbf{-(6 \cdot 20)}$ and $b = a + c = \mathbf{14 = (20 + (-6))}$
The only choice is 20 and -6
Therefore,
$\quad 24x^3y + 42x^2y^2 - 45xy^3$

$= 3xy(8x^2 + 14xy - 15y^2)$

$= 3xy(8x^2 + \mathbf{(20 - 6)}xy - 15y^2)$

$= 3xy[(8x^2 + 20xy) + (-6xy - 15y^2)]$

$= 3xy[4x(2x + 5y) + (-3y)(2x + 5y)]$

$= 3xy(2x + 5y)(4x - 3y)$

◆◆◆◆◆

**EXERCISES

I. Factor the trinomials.

1. $10x^2 - 11x - 6$

              2. $2x^2 + 5x + 6$

3. $6x^2 + xy - 12y^2$

              4. $10a^2 - 17a + 6$

5. $16x^4 + 31x^2 - 2$

              6. $4x^3 + 20x^2 - 56x$

7. $9a^3b + 24a^2b^2 + 16ab^3$

              8. $14a^2 + 3ab - 2b^2$

9. $156a^2 - a - 1$

            10. $15a^2 - 38a + 24$

II. Factor each polynomial completely.

11. $10x^4 - 11x^3 - 6x^2$

            12. $6x^3 + 15x^2 + 18x$

13. $30a^2b - 76ab + 48b$

            14. $36y^4 - 3xy^3 - 18x^2y^2$

15. $24x^3 + 72x^2y + 54xy^2$

            16. $33x^5y^5 - 22x^2y^2$

17. $a^3 - 4a^2 - a + 4$

            18. $x^6 + x^4 + x^3 + x$

19. $3xy - 4y + 3xz - 4z$

            20. $7a + 7b + ma + mb$

◆◆◆◆◆

70

**SOLUTIONS TO EXERCISES

I. Factor the trinomials.

1. $10x^2 - 11x - 6 = (2x - 3)(5x + 2)$

2. $2x^2 + 5x + 6$ There are no numbers to satisfy $ac = 2 \cdot 6 = 12 = 2 \cdot 6$ $= 3 \cdot 4$ and $b = a + c = 5$. Therefore, this is a prime factor.

3. $6x^2 + xy - 12y^2$
   $= (2x + 3y)(3x - 4y)$

4. $10a^2 - 17a + 6$
   $= (2a - 1)(5a - 6)$

5. $16x^4 + 31x^2 - 2$
   $= (16x^2 - 1)(x^2 + 2)$
   $= (4x+1)(4x-1)(x^2 + 2)$

6. $4x^3 + 20x^2 - 56x$
   $= 4x(x^2 + 5x - 14)$
   $= 4x(x + 7)(x - 2)$

7. $9a^3b + 24a^2b^2 + 16ab^3$
   $= ab(9a^2 + 24ab + 16b^2)$
   $= ab(3a + 4b)^2$

8. $14a^2 + 3ab - 2b^2$
   $= (2a + b)(7a - 2b)$

9. $156a^2 - a - 1$
   $= (13a + 1)(12a - 1)$
   $* \ a \cdot c = 156 \cdot (-1) = (12) \cdot (-13)$
   and $a + c = 12 + (-13) = -1$

10. $15a^2 - 38a + 24$
    $= (3a - 4)(5a - 6)$

II. Factor each polynomial completely.

11. $10x^4 - 11x^3 - 6x^2$
    $= x^2(10x^2 - 11x - 6)$
    $= x^2(2x - 3)(5x + 2)$

12. $6x^3 + 15x^2 + 18x$
    $= 3x(2x^2 + 5x + 6)$

13. $30a^2b - 76ab + 48b$
    $= 2b(15a^2 - 38a + 24)$
    $= 2b(3a - 4)(5a - 6)$

14. $36y^4 - 3xy^3 - 18x^2y^2$
    $= 3y^2(12y^2 - xy - 6x^2)$
    $= 3y^2(3y + 2x)(4y - 3x)$

15. $24x^3 + 72x^2y + 54xy^2$
    $= 6x(4x^2 + 12xy + 9y^2)$
    $= 6x(2x + 3y)^2$

16. $33x^5y^5 - 22x^2y^2$
    $= 11x^2y^2(3x^3y^3 - 2)$

17. $a^3 - 4a^2 - a + 4$
    $= a^2(a - 4) - 1 \cdot (a - 4)$
    $= (a - 4)(a^2 - 1)$
    $= (a - 4)(a + 1)(a - 1)$

18. $x^6 + x^4 + x^3 + x$
    $= x^4(x^2 + 1) + x(x^2 + 1)$
    $= (x^2 + 1)(x^4 + x)$
    $= x(x^2 + 1)(x^3 + 1)$

19. $3xy - 4y + 3xz - 4z$
    $= y(3x - 4) + z(3x - 4)$
    $= (3x - 4)(y + z)$

20. $7a + 7b + ma + mb$
    $= 7(a + b) + m(a + b)$
    $= (a + b)(7 + m)$

◆◆◆◆◆

4.5 **MORE FACTORING**

OBJECTIVES:

*(A) To use a factoring strategy.
*(B) To use division to simplify the Factoring.

** **REVIEW**

*(A) Factor strategy.

Step 1. Factor out the common factor if there is one.
Step 2. If the factor is a binomial, check to see if it is the difference of two squares.
Step 3. If the factor is a trinomial, then check to see if it is a perfect square $(a \pm b)^2 = a^2 \pm 2ab + b^2$.
Step 4. Use the ac method and factor by grouping.

Examples:

Factor the following polynomials:

a. $3a^4 - 147a^2$ b. $11x^3y^2 - 66x^2y^3 + 99xy^4$

Solutions:

a. $3a^4 - 147a^2$

Step 1.   Factor out the common factor $3a^2(a^2 - 49)$.
Step 2.   Consider the binomial $(a^2 - 49)$.
          It is the difference of two squares.
          Therefore, $3a^2(a^2 - 49) = 3a^2(a + 7)(a - 7)$

b. $11x^3y^2 - 66x^2y^3 + 99xy^4$

Step 1.   Factor out the common factor $11xy^2(x^2 - 6xy + 9y^2)$.
Step 2.   Consider the trinomial $(x^2 - 6xy + 9y^2) = (x - 3y)^2$.
          This is a perfect square.
          Therefore, $11x^3y^2 - 66x^2y^3 + 99xy^4 = 11xy^2(x - 3y)^2$.

*(B) Use division to simplify factoring.
     Dividing the polynomial by the known factor will simplify the polynomial to a lesser degree polynomial.

Examples:
Factor the polynomial $x^3 - 6x^2 - 7x + 48$, given that the binomial $x - 3$ is a factor of the polynomial.

$$x^2 - 3x - 16$$

$$x - 3 \overline{) x^3 - 6x^2 - 7x + 48}$$
$$\underline{x^3 - 3x^2}$$
$$-3x^2 - 7x + 48$$
$$\underline{-3x^2 + 9x}$$
$$-16x + 48$$
$$\underline{-16x + 48}$$
$$0$$

$$x^3 - 6x^2 - 7x + 48$$
$$= (x - 3)(x^2 - 3x - 16)$$

♦♦♦♦♦

EXERCISES

I. True or False ?

1. $x^2 + 4y^2$ is a difference of two squares.

2. $x^2 + 3x + 4$ is a prime polynomial.

3. $x^2 - 3x - 4$ is a prime polynomial.

4. The polynomial can be factored by grouping.

5. $x^2 - 16 = (x + 4)^2$ for any value of x.

II. Factor each polynomial completely

6. $3m^3n - 3mn^3$

7. $16x^2 - 8x + 1$

8. $16x^2 - 121y^2$

9. $2x^3 - 7x^2 - 98x + 343$

10. $81x^4 - y^2$

11. $4x^2y^2 - 82xy^3 + 40y^4$

12. $m^2a - ma - 156a$

13. $3x^2 + 3x + 3$

III. Factor each polynomial completely, given that the binomial following it is a factor of the polynomial.

14. $x^3 + 2x^2 - x - 2$, $(x + 1)$

15. $27x^3 - 8y^3$, $(3x - 2y)$

16. $a^3 + b^3$, $(a + b)$

17. $4x^3 - x$, $(2x - 1)$

18. $a^3 - 2a^2 - 7a - 4$, $(a - 1)$

19. $36y^4 - 3xy^3 - 18x^2y^2$, $(4y - 3x)$

20. $6x^4 + 21x^3 + 33x^2 + 18x$, $(x + 1)$

♦♦♦♦♦

73

SOLUTIONS TO EXERCISES

I. True or False ?

1. False.   ($x^2 - 4y^2$ is a difference of two squares)

2. True.

3. False. $(x^2 - 3x - 4) = (x - 4)(x + 1)$ is not a prime polynomial

4. True.

5. False. $(x^2 - 16) = (x + 4)(x - 4)$ for any value of x.

II. Factor each polynomial completely

6. $3m^3n - 3mn^3$

   $= 3mn(m^2 - n^2)$

   $= 3mn(m + n)(m - n)$

7. $16x^2 - 8x + 1$

   $= (4x - 1)^2$

8. $16x^2 - 121y^2$

   $= (4x + 11y)(4x - 11y)$

9. $2x^3 - 7x^2 - 98x + 343$

   $= x^2(2x - 7) - 49(2x - 7)$

   $= (2x - 7)(x^2 - 49)$

   $= (2x - 7)(x + 7)(x - 7)$

10. $81x^4 - y^2$

    $= (9x^2 - y)(9x^2 + y)$

11. $4x^2y^2 - 82xy^3 + 40y^4$

    $= 2y^2(2x^2 - 41xy + 20y^2)$

    $= 2y^2(2x - y)(x - 20y)$

12. $m^2a - ma - 156a$

    $= a(m^2 - m - 156)$

    $= a(m - 13)(m + 12)$

13. $3x^2 + 3x + 3$

$\quad = 3(x^2 + x + 1)$,

III. Factor each polynomial completely, given that the binomial
following it is a factor of the polynomial.

14. $x^3 + 2x^2 - x - 2$, $(x + 1)$

$$
\begin{array}{r}
x^2 + x - 2 \\
x + 1 \overline{\smash{)}\ x^3 + 2x^2 - x - 2} \\
\underline{x^3 + x^2} \\
x^2 - x - 2 \\
\underline{x^2 + x} \\
-2x - 2 \\
\underline{-2x - 2} \\
0
\end{array}
$$

$x^3 + 2x^2 - x - 2$
$= (x + 1)(x^2 + x - 2)$

15. $27x^3 - 8y^3$, $(3x - 2y)$

$$
\begin{array}{r}
9x^2 + 6xy + 4y^2 \\
3x - 2y \overline{\smash{)}\ 27x^3 - 8y^3} \\
\underline{27x^3 - 18x^2y} \\
18x^2y - 8y^3 \\
\underline{18x^2y - 12xy^2} \\
12xy^2 - 8y^3 \\
\underline{12xy^2 - 8y^3} \\
0
\end{array}
$$

$27x^3 - 8y^3$
$= (3x - 2y)(9x^2 + 6xy + 4y^2)$

16. $a^3 + b^3$, $(a + b)$

$$
\begin{array}{r}
a^2 - ab + b^2 \\
a + b \overline{\smash{)}\ a^3 + b^3} \\
\underline{a^3 + a^2b} \\
-a^2b + b^3 \\
\underline{-a^2b - ab^2} \\
ab^2 + b^3 \\
\underline{ab^2 + b^3} \\
0
\end{array}
$$

$a^3 + b^3$
$= (a + b)(a^2 - ab + b^2)$

17. $4x^3 - x$, $(2x - 1)$

$$
\begin{array}{r}
2x^2 + x \\
2x - 1 \overline{\smash{)}\ 4x^3 - x} \\
\underline{4x^3 - 2x^2} \\
2x^2 - x \\
\underline{2x^2 - x} \\
0
\end{array}
$$

$4x^3 - x$
$= x(2x - 1)(2x + 1)$

18. $a^3 - 2a^2 - 7a - 4$, $(a + 1)$

$$
\begin{array}{r}
a^2 - 3a - 4 \\
a + 1 \enclose{longdiv}{a^3 - 2a^2 - 7a - 4} \\
\underline{a^3 + a^2} \\
-3a^2 - 7a - 4 \\
\underline{-3a^2 - 3a} \\
-4a - 4 \\
\underline{-4a - 4} \\
0
\end{array}
$$

$a^3 - 2a^2 - 7a - 4$
$= (a + 1)(a^2 - 3a - 4)$
$= (a + 1)(a + 1)(a - 4)$

19. $36y^4 - 3xy^3 - 18x^2y^2$, $(4y - 3x)$

$$
\begin{array}{r}
9y^3 + 6xy^2 \\
4y - 3x \enclose{longdiv}{36y^4 - 3xy^3 - 18x^2y^2} \\
\underline{36y^4 - 27xy^3} \\
24xy^3 - 18x^2y^2 \\
\underline{24xy^3 - 18x^2y^2} \\
0
\end{array}
$$

$36y^4 - 3xy^3 - 18x^2y^2$
$= 3y^2(12y^2 - xy - 6x^2)$
$= 3y^2(4y - 3x)(3y + 2x)$

20. $6x^4 + 21x^3 + 33x^2 + 18x$, $(x + 1)$

$$
\begin{array}{r}
6x^3 + 15x^2 + 18x \\
x + 1 \enclose{longdiv}{6x^4 + 21x^3 + 33x^2 + 18x} \\
\underline{6x^4 + 6x^3} \\
15x^3 + 33x^2 + 18x \\
\underline{15x^3 + 15x^2} \\
18x^2 + 18x \\
\underline{18x^2 + 18x} \\
0
\end{array}
$$

$6x^4 + 21x^3 + 33x^2 + 18x$
$= (x + 1)(6x^3 + 15x^2 + 18x)$
$= (x + 1)(3x)(2x^2 + 5x + 6)$
$= 3x(x + 1)(2x^2 + 5x + 6)$

♦♦♦♦♦

76

4.6 <u>USING FACTORING TO SOLVE EQUATIONS</u>

<u>OBJECTIVES</u>:

*(A) To learn the zero factor property and its applications.

** <u>REVIEW</u>

*(A) The zero factor property and its applications.

Example:

Solve the equation:
$$x^2 - 3x + 2 = 0$$
$$(x - 1)(x - 2) = 0 \quad \longrightarrow \text{ factor the left side of the equation}$$
$(x - 1) = 0$ or $(x - 2) = 0$ --> the product of two factors is
zero therefore, one or the
other factor must be zero.
If $(x - 1) = 0$, then $x = 1$ and If $(x - 2) = 0$, then $x = 2$.
$x = 1$ or $x = 2$

  Check:
    when $x = 1$, $x^2 - 3x + 2 = (1)^2 - 3(1) + 2 = 0$
    when $x = 2$, $x^2 - 3x + 2 = (2)^2 - 3(2) + 2 = 0$

Example:
  The perimeter of a rectangle is 14 feet and the area is 12
  square feet, find the width and the length of the rectangle.

Solution:
  Let the variable **x** represent the unknown quantity (width).
  The perimeter of the rectangle is 14 feet. Therefore, the length is
  $(14 - 2x)/2 = 7 - x$.
  The area = length·width
  $x \cdot (7 - x) = 12$
  $(7x - x^2 - 12) = 0$
  $x^2 - 7x + 12 = 0$ --> multiply by (-1) on both sides
  $(x - 3)(x - 4) = 0$
  $x = 3$ or $x = 4$ --> If the width is 3 ft, then
                         the length is $7 - x = 7 - 3 = 4$ ft
If the width is 4 ft, then the length is $7 - x = 7 - 4 = 3$

  Check:
    The area = $3 \cdot 4 = 12$ and the perimeter = $2 \cdot 3 + 4 \cdot 2 = 14$

◆◆◆◆◆

I. Solve each equation.

1. $(x - 1)(x + 5) = 0$

2. $(x - 0)(x - 5) = 0$

3. $(x + 2)^2 = 0$

4. $(3x + 2)(4x + 3) = 0$

5. $3m^2 + 6m = 0$

6. $4x^2 + 4x = -1$

7. $x^2 - 9 = 0$

8. $3x^2 - 3x = 0$

9. $a^4 = a^3$

10. $x^2 + x - 56 = 0$

11. $3m^3 - 5m^2 - 27m + 45 = 0$

12. $3 - x - x^2 = 0$

13. $9x^2 = x$

14. $x^2 - 11x + 30 = 0$

15. $4x^3 - 12x^2 - 16x + 48 = 0$

16. $x^2 - 7/3x - 20/3 = 0$

II. Solve each problem.

17. The product of three consecutive positive integers is -6.
    Find these numbers.
18. The width of a rectangle is three times the length.
    The area of the rectangle is 48. Find the length and the width.
19. The sum of two integers is 2 and their product is -35.
    Find the integers.
20. One child is three times older than the second child.
    Two years ago, the product of their ages was 20.
    Find the present ages of the children.

♦♦♦♦♦

SOLUTIONS TO EXERCISES

I. Solve each equation.

1. $(x - 1)(x + 5) = 0$
    $x = 1$ or $x = -5$

2. $(x - 0)(x - 5) = 0$
    $x = 0$ or $x = 5$

3. $(x + 2)^2 = 0$
    $x = -2$

4. $(3x + 2)(4x + 3) = 0$
    $x = -2/3$ or $x = -3/4$

5. $3m^2 + 6m = 0$
    $3m(m + 2) = 0$,
    $m = 0$ or $m = -2$

6. $4x^2 + 4x = -1$
    $4x^2 + 4x + 1 = 0$
    $(2x + 1)^2 = 0$
    $x = -1/2$

7. $x^2 - 9 = 0$
    $(x + 3)(x - 3) = 0$
    $x = -3$ or $x = 3$

8. $3x^2 - 3x = 0$
    $3x(x - 1) = 0$
    $x = 0$ or $x = 1$

78

9. $a^4 = a^3$
$a^4 - a^3 = 0$
$a^3(a - 1) = 0$
$a = 0$ or $a = 1$

10. $x^2 + x - 56 = 0$
$(x + 8)(x - 7) = 0$
$x = -8$ or $x = 7$

11. $3m^3 - 5m^2 - 27m + 45 = 0$
$m^2(3m - 5) - 9(3m - 5)$
$(3m - 5)(m^2 - 9) = 0$
$(3m - 5)(m - 3)(m + 3) = 0$
$m = 5/3$ or $m = 3$ or $m = -3$

12. $3 - x - x^2 = 0$
$-x^2 - x + 3 = 0$
$(x^2 + x - 3) = 0$
This is a prime polynomial

13. $9x^2 = x$
$9x^2 - x = 0$
$x(9x - 1) = 0$
$x = 0$ or $x = 1/9$

14. $x^2 - 11x + 30 = 0$
$(x - 6)(x - 5) = 0$
$x = 6$ or $x = 5$

15. $4x^3 - 12x^2 - 16x + 48 = 0$
$4x^2(x - 3) - 16(x - 3) = 0$
$(x - 3)(4x^2 - 16) = 0$
$4(x - 3)(x - 2)(x + 2) = 0$
$x = 3$, $x = 2$, $x = -2$

16. $x^2 - (7/3)x - (20/3) = 0$
$3x^2 - 7x - 20 = 0$
$(3x + 5)(x - 4) = 0$
$x = -5/3$ or $x = 4$

II. Solve each problem.

17. The product of three consecutive positive integers is -6.
Find these numbers.

Let x be one integer and the other two be x + 1 and x + 2.
$x(x + 1)(x + 2) = -6$ --> multiply it out
$x^3 + 3x^2 + 2x + 6 = 0$ --> factor out the common factor
$x^2(x + 3) + 2(x + 3) = 0$
$(x + 3)(x^2 + 2) = 0$
$x = -3$ or $x^2 = -2$ (complex numbers)
Therefore, $x = -3$, $x + 1 = -2$, and $x + 2 = -1$
Three integers are -3, -2, and -1.
Check:
The product $(-3)(-2)(-1) = -6$.

18. The width of a rectangle is three times the length.
The area of the rectangle is 48. Find the length and the width.

Let x be the length and the width be $(3 \cdot \text{length}) = 3x$.
The area $= x \cdot 3x = 48$
$3x^2 = 48$
$x^2 - 16 = 0$
$(x - 4)(x + 4) = 0$
$x = 4$ or $x = -4$ (impossible)
Therefore the length is 4 feet and the width is 12 feet.
Check:
area $= 4 \cdot 12 = 48$

79

19. The sum of two integers is 2 and their product is -35.
    Find the integers.

    Let x be one of the integers, x + (other integer) = 2.
    Therefore the other integer is equal to 2 - x.
    $x \cdot (2 - x) = -35$ --> the product of two integers is -35
    $2x - x^2 + 35 = 0$
    $x^2 - 2x - 35 = 0$
    $(x - 7)(x + 5) = 0$
    $x = 7$ or $x = -5$
    The numbers are 7 and -5
    Check:
    $(7) + (-5) = 2$ and $7 \cdot (-5) = -35$

20. One child is three times older than the second child.
    Two years ago, the product of their ages was 20.
    Find the present ages of the children.

    Let x be the age of one child and the other child's age be 3x,
    $(x - 2)(3x - 2) = 20$ --> two years ago the product of
                                    their ages is 20.
    $(x - 2)(3x - 2) = 20$
    $3x^2 - 8x + 4 - 20 = 0$
    $3x^2 - 8x - 16 = 0$
    $(3x + 4)(x - 4) = 0$ ---> ($x = -4/3$ years old is impossible)
    $x = 4$ and $3x = 12$.
    Therefore, their ages are 4 and 12.
    Check:
    3 times $x = 3 \cdot 4 = 12$ and $(4 - 2) \cdot (12 - 2) = 20$

◆◆◆◆◆

CHAPTER IV - TEST A

I. Find the prime factorization for each integer

   1. 64   2. 63   3. 720
   4. 231   5. 550   6. 63

II. Find the greatest common factor for each group of integers

   7. 41, 17, 13   8. 9, 90, 900   9. 12, 144, 60
   10. 38, 57, 76   11. 34, 51, 68   12. 32, 64, 256

III. Complete the factorization by filling in the parentheses

   13. $20x - 5x^2 = ($ $)(x - 4)$

   14. $m^3 + 9m^2 - 22m = m($ $)(m - 2)$

   15. $9xy^2 + 90y^3 = ($ $)(x + 10y)$

16. $-11x^2 - 22x - 11 = -11(\quad)^2$

17. $3x + 5y + 9xy + 15y^2 = (\quad)(1 + 3y)$

18. $169m^4 - 100n^2 = (13m^2 - 10n)(\quad)$

19. $-a^2 - a + 1 = (-1)(\quad)$

20. $100m^2 - 10000n^2 = 100(\quad)(\quad)$

◆◆◆◆◆

CHAPTER IV - TEST B

I. Factor each polynomial completely

1. $1 - x^2y^4$ 2. $a^4 - b^4$

3. $-4 - 4x - x^2$ 4. $a^4 - a^2$

5. $x^2 - x - 30$ 6. $cy - bc + y^2 - by$

7. $3x^3 + 3x^2y + 3xy^2 + 3y^3$ 8. $30x^2 + 11xy - 30y^2$

9. $6x^2 + 9xy - 4xz - 6yz$ 10. $2x - x^2 - 1$

11. $m - 3m^2 + x^3 - 3x^3m$ 12. $9x^3 - 25x$

II. Factor each polynomial completely, given that the binomial following it is a factor of the polynomial.

13. $x^3 + x^2 + x + 1$, $(x+1)$ 14. $27x^3 - y^3$, $(3x-y)$

15. $3x^3 + 8x^2 + 8x + 5$, $(3x+5)$ 16. $3x^2 + 8x - 11$, $(x-1)$

III. Solve each equation

17. $x(x - 1)(x - 2) = 0$ 18. $3(x^2 - 9) = 0$

19. $21x^2 - 58x + 21 = 0$ 20. $6x^2 + 13x + 5 = 0$

◆◆◆◆◆

CHAPTER IV - TEST C

I. Factor each polynomial completely

1. $3x^3 + 18x^2 - 81x$ 2. $x^2 - 4xy - 21y^2$

3. $a^2 + 10a + 25$ 4. $-11a^3 + 88a^2b + 363ab^2$

5. $-3ab^2 - 9ab + 12a$ 6. $x^4 - 9x^2 - 36$

7. $6x^3 + 6x^2 - 72x$

8. $5x^4 + 70x^3 + 225x^2$

9. $4y^3 + 36y^2 + 72y$

10. $y^2 + 85y + 36$

11. $-2y^4 + 26y^3 + 96y^2$

12. $x^4 - 22x^3 + 12x^2$

13. $a^2b - 15ab + 36b$

14. $45x^2y^2 + 14a^2b + 45a^2$

15. $30 + m^2 - 11m$

16. $-x^2 + 5x + 24$

III. Solve the following problems
   (● Let x be the unknown quantity)
   (● Express the problem as an algebraic expression)

17. The sum of two integers is 4. The sum of the squares of the two integers is 58. Find the numbers.

18. The sum of two integers is 32. Their product is 220. Find the numbers.

19. The length of each side of a square is cut 3 inches. The area of the resulting square is 36 in square. Find the length of a side of the original square.

20. The product of two consecutive even positive integers is 120. Find the numbers.

♦♦♦♦♦

CHAPTER IV - TEST A - SOLUTIONS

I. Find the prime factorization for each integer

1. $64 = 2^6$

2. $63 = 3^2 \cdot 7$

3. $720 = 2^4 \cdot 3^2 \cdot 5$

4. $231 = 3 \cdot 7 \cdot 11$

5. $550 = 2 \cdot 5^2 \cdot 11$

6. $63 = 3^2 \cdot 7$

II. Find the greatest common factor for each group of integers

7. 41,17,13   (GCF=1)

8. 9,90,900   (GCF=9)

9. 12,144,60   (GCF=12)

10. 38,57,76   (GCF=19)

11. 34,51,68   (GCF=17)

12. 32,64,256   (GCF=32)

III. Complete the factorization by filling in the parentheses

13. $20x - 5x^2 = (\underline{\mathbf{-5x}})(x - 4)$

14. $m^3 + 9m^2 - 22m = m(\underline{\mathbf{m + 11}})(m - 2)$

15. $9xy^2 + 90y^3 = (\underline{\mathbf{9y^2}})(x + 10y)$

16. $-11x^2 - 22x - 11 = -11(\underline{\mathbf{x + 1}})^2$

17. $3x + 5y + 9xy + 15y^2 = (\underline{\mathbf{3x + 5y}})(1 + 3y)$

18. $169m^4 - 100n^2 = (13m^2 - 10n)\ (\underline{\mathbf{13m^2 + 10n}})$

19. $-a^2 - a + 1 = (-1)(\underline{\mathbf{a^2 + a - 1}})$

20. $100m^2 - 10000n^2 = 100(\underline{\mathbf{m - 10n}})(\underline{\mathbf{m + 10n}})$

♦♦♦♦♦

CHAPTER IV - TEST B - SOLUTIONS

I. Factor each polynomial completely

   1. $1 - x^2y^4 = (1 - xy^2)(1 + xy^2)$

   2. $a^4 - b^4 = (a^2 + b^2)(a + b)(a - b)$

   3. $-4 - 4x - x^2 = -(x^2 + 4x + 4) = -(x + 2)^2$

   4. $a^4 - a^2 = a^2(a + 1)(a - 1)$

   5. $x^2 - x - 30 = (x - 6)(x + 5)$

   6. $cy - bc + y^2 - by = (y - b)(y + c)$

   7. $3x^3 + 3x^2y + 3xy^2 + 3y^3 = 3(x^2 + y^2)(x + y)$

   8. $30x^2 + 11xy - 30y^2 = (6x - 5y)(5x + 6y)$

   9. $6x^2 + 9xy - 4xz - 6yz = (2x + 3y)(3x - 2z)$

  10. $2x - x^2 - 1 = -(x - 1)^2$

  11. $m - 3m^2 + x^3 - 3x^3m = (x^3 + m)(1 - 3m)$

  12. $9x^3 - 25x = x(3x - 5)(3x + 5)$

II. Factor each polynomial completely, given that the binomial following it is a factor of the polynomial.

  13. $x^3 + x^2 + x + 1 = (\mathbf{x^2 + 1})(x + 1)$

  14. $27x^3 - y^3 = (\mathbf{9x^2 + 3xy + y^2})(3x - y)$

  15. $3x^3 + 8x^2 + 8x + 5 = (\mathbf{x^2 + x + 1})(3x + 5)$

  16. $3x^2 + 8x - 11 = (\mathbf{3x + 11})(x - 1)$

III. Solve each equation

17. $x(x - 1)(x - 2) = 0$ $(x = 0, x = 1, x = 2)$

18. $3(x^2 - 9) = 0$
    $3(x - 3)(x + 3) = 0$ $(x = 3, x = -3)$

19. $21x^2 - 58x + 21 = 0$
    $(7x - 3)(3x - 7) = 0$ $(x = 3/7, x = 7/3)$

20. $6x^2 + 13x + 5 = 0$
    $(2x + 1)(3x + 5) = 0$ $(x = -1/2, x = -5/3)$

♦♦♦♦♦

CHAPTER IV - TEST C - SOLUTIONS

I. Factor each polynomial completely

1. $3x^3 + 18x^2 - 81x$
   $= 3x(x^2 + 6x - 27)$
   $= 3x(x + 9)(x - 3)$

2. $x^2 - 4xy - 21y^2$
   $= (x + 3y)(x - 7y)$

3. $a^2 + 10a + 25$
   $= (a + 5)^2$

4. $-11a^3 + 88a^2b + 363ab^2$
   $= -11a(a + 3b)(a - 11b)$

5. $3ab^2 + 9ab - 12a$
   $= 3a(b - 1)(b + 4)$

6. $x^4 - 9x^2 - 36$
   $= (x^2 - 12)(x^2 + 3)$

7. $6x^3 + 6x^2 - 72x$
   $= 6x(x - 3)(x + 4)$

8. $5x^4 + 70x^3 + 225x^2$
   $= 5x^2(x + 5)(x + 9)$

9. $4y^3 + 36y^2 + 72y$
   $= 4y(y + 3)(y + 6)$

10. $y^2 + 85y + 36$
    This is a prime polynomial

11. $-2y^4 + 26y^3 + 96y^2$
    $= -2y^2(y + 3)(y - 16)$

12. $x^4 - 22x^3 + 12x^2$
    This is a prime polynomial

13. $a^2b - 15ab + 36b$
    $= b(a - 3)(a - 12)$

14. $45x^2y^2 + 14a^2b + 45a^2$
    This is a prime polynomial

15. $30 + m^2 - 11m$
    $= (m - 6)(m - 5)$

16. $-x^2 + 5x + 24$
    $= -(x^2 - 5x - 24)$
    $= -(x + 3)(x - 8)$

III. Solve the following problems
    (• Let x be the unknown quantity)
    (• Express the problem as an algebraic expression)

17. The sum of two integers is 4. The sum of the squares
    of the two integers is 58. Find the numbers.

    Let x be one integer and the other integer be
    (4 - x) --> Since the sum of two integers is 4.
    x^2 + $(4 - x)^2$ = 58 --> The sum of the squares of the two is 58
    x^2 + 16 - 8x + x^2 = 58
    $2x^2$ - 8x - 42 = 0
    2(x + 3)(x - 7) = 0
    The integers are -3 and 7.

18. The sum of two integers is 32. Their product is 220.
    Find the numbers.

    Let x be one integer and the other integer be (32 - x).
    x· (32 - x) = 220 --> The product of the two is 220.
    x^2 - 32x - 220 = 0
    (x - 22)(x - 10) = 0
    x = 10 or x = 22
    The numbers are 22 and 10.

19. The length of each side of a square is cut 3 inches.
    The area of the resulting square is 36 in square.
    Find the length of a side of the original square.

    If x is the length of the square, then the area is equal to x^2.
    $(x-3)^2$ = 36   each side is cut 3 inches
    x^2 - 6x + 9 - 36 = 0
    (x + 3)(x - 9) = 0
    x = 9. The length of the square is 9 inches.

20. The product of two consecutive even positive integers is 120.
    Find the numbers.

    Let x be one integer, then x· (x + 2) = 120.
    x^2 + 2x - 120 = 0
    (x + 12)(x - 10) = 0
    x = - 12 or x = 10
    The numbers are 10 and 12.

♦♦♦♦♦

CHAPTER 5

RATIONAL EXPRESSIONS

5.1 PROPERTIES OF RATIONAL EXPRESSIONS

OBJECTIVES:

*(A) To define rational numbers and rational expressions.
*(B) To simplify rational numbers and rational expressions.
*(C) To reduce rational numbers with the quotient rule.

** REVIEW

*(A) ● A rational number is the ratio of two integers where the
        denominator is not equal to zero.
      ● A rational expression is the ratio of two polynomials where the
        denominator is not equal to zero.

Examples:

  a. $\frac{3}{4}$, $\frac{(-3)}{5}$, $\frac{3}{1}$, and $\frac{1}{4}$ are rational numbers.

  b. $\frac{x^3-x^2+x+1}{x-2}$, $\frac{2x-5}{x^3+x+1}$, $\frac{8}{-7}$, $5-x = \frac{5-x}{1}$ are rational expressions.

Examples: Which value(s) of x cannot be used in each rational
          expression?

  a. $\frac{x+1}{x^2-1}$,   b. $\frac{3x^2}{x^2+2x+1}$,   c. $\frac{x-2}{3x+5}$

Solutions:
   a. The denominator is 0 if $x^2 - 1 = (x + 1)(x - 1) = 0$.
      Therefore x cannot be equal to +1 or -1.

   b. The denominator is 0 if $x^2 + 2x +1 = (x + 1)^2 = 0$.
      Therefore x cannot be equal to -1.

   c. The denominator is 0 if $3x + 5 = 0$.
      Therefore x cannot be equal to -(5/3).

*(B) Reduce rational numbers to lowest terms.

   ● Factor out the common factors of the denominator and the
     numerator.

   ● Divide the numerator and the denominator by the GCF
     (greatest common factor).

86

Examples:

       a. $\dfrac{35}{147}$ b. $\dfrac{x^3-x^2-12x}{x^2-4x}$

Solutions:

     a. $\dfrac{35}{147} = \dfrac{5 \cdot 7}{3 \cdot 7 \cdot 7} = \dfrac{5}{21}$ $GCF = 7$

     b. $\dfrac{x^3-x^2-12x}{x^2-4x} = \dfrac{x(x^2-x-12)}{x(x-4)} = \dfrac{(x-4)(x+3)}{(x-4)} = x+3$ $GCF = x(x-4)$

*(C) Reduce by the quotient rule.
    Consider:

       $\dfrac{a^m}{a^n} = a^{m-n}, \quad a^{-k} = \dfrac{1}{a^k} \quad for \ \ a \neq 0$

Examples:
    Reduce the following rational expressions to their lowest terms:

     a. $\dfrac{9x^7}{3x}$ b. $\dfrac{39x^2y^3}{26x^3y^4}$

Solutions:

     a. $\dfrac{9x^7}{3x} = \dfrac{3 \cdot 3x^7}{3x} = \dfrac{(3x)(3x^6)}{3x} = 3x^6$

     b. $\dfrac{39x^2y^3}{26x^3y^4} = \dfrac{3 \cdot 13}{2 \cdot 13} \cdot (x^{2-3}y^{3-4}) = \dfrac{3}{2}x^{-1}y^{-1} = \dfrac{3}{2xy}$

    Consider $(y - x) = -(x - y)$
    If you divide $(x - y)$ by $(y - x)$, then

     $\dfrac{(x-y)}{(y-x)} = \dfrac{(x-y)}{-(x-y)} = -1$

Examples:

     a. $\dfrac{3x^2-11x-20}{-x^2+3x+10} = \dfrac{(3x+4)(x-5)}{-(x^2-3x-10)} = \dfrac{(x-5)(3x+4)}{-(x-5)(x+2)} = \dfrac{-(3x+4)}{(x+2)}$

     b. $\dfrac{4a^2-9b^2}{3b-2a} = \dfrac{(2a-3b)(2a+3b)}{3b-2a} = \dfrac{(2a-3b)(2a+3b)}{-(2a-3b)} = -(2a+3b)$

♦♦♦♦♦

I. Which values of x cannot be used in each rational expression?

1. $\dfrac{x-5}{(x-1)(x+2)}$ 2. $\dfrac{1}{x^2-49}$ 3. $\dfrac{x}{3}$ 4. $\dfrac{x+4}{x}$

5. $\dfrac{x}{4x-7}$ 6. $\dfrac{x^2-x+1}{9-2x}$ 7. $\dfrac{x}{x-4}$ 8. $\dfrac{1}{(x-1)(x-2)(x-3)}$

II. Reduce the following rational expressions to lowest terms.

9. $\dfrac{x^9}{x^6}$ 10. $\dfrac{x^{12}}{x^{13}}$ 11. $\dfrac{a^8}{a^3}$

12. $\dfrac{2a^3b^5}{6a^5b^3}$ 13. $\dfrac{-36m^{10}n^7}{-(4m^{13}n^{12})}$ 14. $\dfrac{14x^2y^3z^4}{2xy^3z^2}$

15. $\dfrac{10a^{35}b}{1000ab^{35}}$ 16. $\dfrac{x^2-a^2}{x+a}$ 17. $\dfrac{3x^2+7x+4}{6x^2+17x+12}$

18. $\dfrac{-(x-2)}{x^3-4x}$ 19. $\dfrac{17x^3-85x^2}{17}$ 20. $\dfrac{7x-21x^2}{7x}$

◆◆◆◆◆

SOLUTIONS TO EXERCISES

I. Which values of x cannot be used in each rational expression?
   (Hint: Let the denominator equal to zero and solve for x)

1. $(x - 1)(x + 2) = 0$
   $x = 1, x = -2$

2. $x^2 - 49 = (x + 7)(x - 7) = 0$
   $x = 7, x = -7$

3. x can be any real number

4. $x = 0$

5. $(4x - 7) = 0$
   $x = 7/4$

6. $9 - 2x = 0$
   $x = 9/2$

7. $x - 4 = 0$
   $x = 4$

8. $(x - 1)(x - 2)(x - 3) = 0$
   $x = 1, x = 2, x = 3$

◆◆◆◆◆

II. Reduce the following rational expressions to lowest terms.

9. $\dfrac{x^9}{x^6} = x^{9-6} = x^3$

10. $\dfrac{x^{12}}{x^{13}} = x^{12-13} = x^{-1} = \dfrac{1}{x}$

11. $\dfrac{a^8}{a^3} = a^{8-3} = a^5$

12. $\dfrac{2a^3b^5}{6a^5b^3} = \dfrac{1}{3}(a^{3-5}b^{5-3}) = \dfrac{1}{3}a^{-2}b^2$

$= \dfrac{b^2}{3a^2}$

13. $\dfrac{-36m^{10}n^7}{-(4m^{13}n^{12})} = 9m^{10-13}n^{7-12} = 9m^{-3}n^{-5}$

14. $\dfrac{14x^2y^3z^4}{2xy^3z^2} = 7x^{2-1}y^{3-3}z^{4-2} = 7xz^2$

$= \dfrac{9}{m^3n^5}$

15. $\dfrac{10a^{35}b}{1000ab^{35}} = \dfrac{1}{100}a^{35-1}b^{1-35} = \dfrac{1}{100}a^{34}b^{-34}$

16. $\dfrac{x^2-a^2}{x+a} = \dfrac{(x+a)(x-a)}{x+a} = x-a$

$= \dfrac{a^{34}}{100b^{34}}$

17. $\dfrac{3x^2+7x+4}{6x^2+17x+12} = \dfrac{(x+1)(3x+4)}{(2x+3)(3x+4)} = \dfrac{x+1}{2x+3}$

18. $\dfrac{-(x-2)}{x^3-4x} = \dfrac{-(x-2)}{x(x+2)(x-2)} = \dfrac{-1}{x(x+2)}$

19. $\dfrac{17x^3-85x^2}{17} = \dfrac{17x^2(x-5)}{17} = x^2(x-5)$

20. $\dfrac{7x-21x^2}{7x} = \dfrac{7x(1-3x)}{7x} = 1-3x$

◆◆◆◆◆

5.2 **MULTIPLICATION AND DIVISION**

OBJECTIVES:

*(A) To multiply rational numbers and rational expressions.
*(B) To divide rational numbers and rational expressions.

** **REVIEW**

*(A) Multiply rational numbers or rational expressions:

1. Multiply the numerator. The product is the new numerator.
2. Multiply the denominator. The product is the new denominator.
3. Factor out the common factors of the numerator and denominator.

Examples:

a. $\dfrac{7}{12} \cdot \dfrac{5}{3} = \dfrac{7 \cdot 5}{12 \cdot 3} = \dfrac{35}{36}$

b. $\dfrac{x-2}{5} \cdot \dfrac{x}{9x^2-36} = \dfrac{x \cdot (x-2)}{5 \cdot 9 \cdot (x^2-4)} = \dfrac{x(x-2)}{45(x-2)(x+2)} = \dfrac{x}{45(x+2)}$

The factor is $(x - 2)$ where $x \neq \pm 2$

*(B) ● Divide rational numbers:
Invert the second rational number and
multiply the two rational numbers.

   ● Divide rational expressions:
Invert the second rational expression and
multiply the two rational expressions.

Examples:

a. $\dfrac{7}{8} \div \dfrac{15}{3} = \dfrac{7}{8} \times \dfrac{3}{15} = \dfrac{7 \cdot 3}{8 \cdot 15} = \dfrac{21}{120} = \dfrac{7}{40}$

b. $\dfrac{15x}{9} \div \dfrac{3x}{27} = \dfrac{15x}{9} \times \dfrac{27}{3x} = \dfrac{3 \cdot 5 \cdot 3 \cdot 3 \cdot 3 \cdot x}{3 \cdot 3 \cdot 3 \cdot x} = \dfrac{15x}{x} = 15x^{(1-1)} = 15x^0 = 15 \times 1 = 15$

c. $\dfrac{3x^2+6x+3}{3x+1} \div \dfrac{x+1}{x^2+1} = \dfrac{3(x^2+2x+1)}{3x+1} \times \dfrac{x^2+1}{x+1} = \dfrac{3(x+1)^2}{3x+1} \times \dfrac{x^2+1}{x+1} = \dfrac{3(x+1)(x^2+1)}{3x+1}$

d. $\dfrac{x+\dfrac{1}{x}}{x-\dfrac{1}{x}} = (x+\dfrac{1}{x}) \div (x-\dfrac{1}{x}) = (\dfrac{x^2+1}{x}) \div (\dfrac{x^2-1}{x}) = (\dfrac{x^2+1}{x}) \times (\dfrac{x}{x^2-1}) = \dfrac{x^2+1}{x^2-1}$

◆◆◆◆◆

EXERCISES

I. True or False ?

1. $\dfrac{-420}{-210} = 2$

2. $\dfrac{x-4}{2} \times \dfrac{-20}{x-4} = -10$ *for* $x \neq 4$

3. $\dfrac{x-y}{y-x} \div \dfrac{y-x}{y-x} = -1$

4. $\dfrac{\frac{m^2-5}{2}}{5} = \dfrac{m^2-5}{2} \times \dfrac{1}{5}$

5. $\dfrac{3}{a-1} \div \dfrac{1-a}{3} = -1$ *for* $a \neq 1$

6. $\dfrac{3-x}{2} \div (8x+5) = \dfrac{2}{3-x} \times (8x+5)$

II Perform the indicated operations.

7. $\dfrac{3}{5} \times \dfrac{25}{-12}$

8. $\dfrac{91}{12} \div \dfrac{7}{4}$

9. $\dfrac{3x+5}{3} \div \dfrac{9x^2+30x+25}{5}$

10. $\dfrac{9x^2-4y^2}{9} \div \dfrac{3x+2y}{27}$

11. $(2x-y) \div \dfrac{4x^2-y^2}{y}$

12. $\dfrac{35}{x^2-4} \div \dfrac{7xy}{x-2}$

13. $\dfrac{4-x^2}{x^4+5x^3} \div \dfrac{2x^2+x-10}{x^3+4x^2-5x}$

14. $\dfrac{a+3ab}{4a^2} \div \dfrac{3b+1}{2a^2}$

15. $\dfrac{\frac{2x^2+5x+2}{2x+1}}{7}$

16. $\dfrac{\frac{a+b}{19a^2}}{\frac{a-b}{3a}}$

17. $\dfrac{\frac{3}{5} \div \frac{9}{10}}{\frac{8}{3} \times \frac{3}{9}}$

18. $\dfrac{2x}{\frac{x}{2x+3}}$

19. $\dfrac{\frac{x-y}{4x^3}}{\frac{x+y}{x}}$

20. $5x \div \dfrac{1+\frac{2}{x}}{\frac{x+2}{x-1}}$

◆◆◆◆◆

91

SOLUTIONS TO EXERCISES

I True or False ?

   1. True.   2. True.   3. True.   4. True.

   5. *False.* $\dfrac{3}{a-1} \div \dfrac{1-a}{3} = \dfrac{3}{a-1} \times \dfrac{3}{1-a} = \dfrac{-9}{(1-a)^2}$

   6. *False.* $\dfrac{3-x}{2} \div (8x+5) = \dfrac{3-x}{2} \times \dfrac{1}{8x+5} = \dfrac{3-x}{2(8x+5)}$

II Perform the indicated operation.

   7. $\dfrac{3}{5} \times \dfrac{25}{-12} = \dfrac{3 \cdot 5 \cdot 5}{-5 \cdot 2 \cdot 2 \cdot 3} = \dfrac{-5}{4}$

   8. $\dfrac{91}{12} \times \dfrac{4}{7} = \dfrac{7 \cdot 13 \cdot 2 \cdot 2}{3 \cdot 2 \cdot 2 \cdot 7} = \dfrac{13}{3}$

   9. $\dfrac{3x+5}{3} \times \dfrac{5}{9x^2+30x+25} = \dfrac{5(3x+5)}{3(3x+5)^2} = \dfrac{5}{3(3x+5)}$

  10. $\dfrac{9x^2-4y^2}{9} \times \dfrac{27}{3x+2y} = \dfrac{27(3x+2y)(3x-2y)}{9(3x+2y)} = 3(3x-2y)$

  11. $(2x-y) \times \dfrac{y}{4x^2-y^2} = (2x-y) \times \dfrac{y}{(2x-y)(2x+y)} = \dfrac{y}{2x+y}$

  12. $\dfrac{35}{x^2-4} \times \dfrac{x-2}{7xy} = \dfrac{35(x-2)}{7xy(x-2)(x+2)} = \dfrac{5}{xy(x+2)}$

  13. $\dfrac{4-x^2}{x^4+5x^3} \times \dfrac{x^3+4x^2-5x}{2x^2+x-10} = \dfrac{-(-2+x)(2+x)}{x^3(x+5)} \times \dfrac{x(x+5)(x-1)}{(2x+5)(x-2)} = \dfrac{-(x-1)(x+2)}{x^2(2x+5)}$

  14. $\dfrac{a+3ab}{4a^2} \times \dfrac{2a^2}{3b+1} = \dfrac{a(1+3b)(2a^2)}{(4a^2)(3b+1)} = \dfrac{a}{2}$

  15. $(2x^2+5x+2) \div \dfrac{(2x+1)}{7} = \dfrac{(2x+1)(x+2)}{1} \times \dfrac{7}{(2x+1)} = 7(x+2)$

♦♦♦♦♦

92

16. $\dfrac{a+b}{19a^2} \div \dfrac{a-b}{3a} = \dfrac{a+b}{19a^2} \cdot \dfrac{3a}{a-b} = \dfrac{3(a+b)}{19a(a-b)}$

17. $\dfrac{\dfrac{3}{5} \div \dfrac{9}{10}}{\dfrac{8}{3} \cdot \dfrac{3}{9}} = \dfrac{\dfrac{3}{5} \cdot \dfrac{10}{9}}{\dfrac{8}{3} \cdot \dfrac{3}{9}} = \dfrac{\dfrac{2}{3}}{\dfrac{8}{9}} = \dfrac{2}{3} \cdot \dfrac{9}{8} = \dfrac{3}{4}$

18. $\dfrac{2x}{\dfrac{x}{2x+3}} = 2x \div \dfrac{x}{2x+3} = (2x) \cdot \dfrac{2x+3}{x} = 2(2x+3)$

19. $\dfrac{\dfrac{x-y}{4x^3}}{\dfrac{x+y}{x}} = \dfrac{x-y}{4x^3} \div \dfrac{x+y}{x} = \dfrac{x-y}{4x^3} \cdot \dfrac{x}{x+y} = \dfrac{x-y}{4x^2(x+y)}$

20. $(5x) \div \dfrac{\dfrac{x+2}{x}}{\dfrac{x+2}{x-1}} = 5x \div \left(\dfrac{x+2}{x} \cdot \dfrac{x-1}{x+2} \right) = (5x) \div \dfrac{x-1}{x} = (5x) \cdot \dfrac{x}{x-1} = \dfrac{5x^2}{x-1}$

♦♦♦♦♦

5.3 BUILDING UP THE DENOMINATOR

OBJECTIVES:

*(A) To change the denominator.
*(B) To find the LCD for rational numbers and rational expressions.

** REVIEW

*(A) To change the denominator, multiply both the numerator and the denominator by the same factor.

93

Examples:

a. $\dfrac{3}{x+1} = \dfrac{3 \cdot (x+2)}{(x+1) \cdot (x+2)} = \dfrac{3x+6}{x^2+3x+2} = \dfrac{(3x+6) \cdot x}{(x^2+3x+2) \cdot x} = \dfrac{3x^2+6x}{x^3+3x+2x}$

b. $\dfrac{7}{22} = \dfrac{7 \times 9}{22 \times 9} = \dfrac{63}{198} = \dfrac{63 \times \frac{1}{3}}{198 \times \frac{1}{3}} = \dfrac{21}{66}$

c. $\dfrac{a+b}{a-b} = \dfrac{(a+b) \cdot x}{(a-b) \cdot x} = \dfrac{ax+bx}{ax-bx} = \dfrac{(ax+bx)(4x^2)}{(ax-bx)(4x^2)} = \dfrac{4ax^3+4bx^3}{4ax^3-4bx^3}$

*(B) To find the LCD for the denominators remember:

Step 1. Factor each denominator completely using exponential notation.
Step 2. LCD = The product of all the different factors with the highest exponent.

Examples:

a. Find the LCD for the denominators 36 and 48.

$36 = 2^2 \cdot 3^2$ and $48 = 2^4 \cdot 3^1$

The highest power of 2 is 4

The highest power of 3 is 2

LCD $= 2^4 \cdot 3^2 = 144$

b. Find the LCD for the denominators $a^2 + 4a + 3$ and $a^2 + 6a + 9$.

$a^2 + 4a + 3 = (a + 3)(a + 1)$

$a^2 + 6a + 9 = (a + 3)^2$

The highest power of $(a + 3)$ is 2

The highest power of $(a + 1)$ is 1

LCD $= (a + 3)^2 \cdot (a + 1)$

c. Find the LCD of the rational expressions

$a^2 - 2a - 3$ and $a^2 + 2a + 1$

$a^2 - 2a - 3 = (a - 3)(a + 1)$

$a^2 + 2a + 1 = (a + 1)^2$

LCD $= (a - 3)(a + 1)^2$

◆◆◆◆◆

EXERCISES

I. Convert each rational expression into an equivalent rational expression with the indicated denominator or numerator.

1. $\dfrac{1}{4} = \dfrac{?}{40}$

2. $\dfrac{-2}{5} = \dfrac{12}{?}$

3. $a = \dfrac{1}{?}$

4. $\dfrac{7}{x^2 y} = \dfrac{?}{x^3 y^3}$

5. $\dfrac{x+3}{2x+5} = \dfrac{?}{2x^2+7x+5}$

6. $\dfrac{5}{9-x^2} = \dfrac{5(x+1)}{?}$

7. $17 = \dfrac{17a^2}{?}$

8. $-3 = \dfrac{?}{2x+1}$

9. $\dfrac{x-y}{y-x} = \dfrac{?}{x-y}$

10. $\dfrac{7x+9}{7x^2+9x} = \dfrac{1}{?}$

11. $\dfrac{x-4}{6} = \dfrac{x^2-16}{?}$

12. $\dfrac{3x}{-5y^2} = \dfrac{?}{-15y^2}$

II. Find the LCD for each pair of rational expressions, and convert each rational expression into an equivalent rational expression with the LCD as the denominator.

13. $\dfrac{7}{51}$, $\dfrac{3}{68}$

14. $\dfrac{3}{5ab^2c}$, $\dfrac{4}{7a^2bc}$

15. $\dfrac{9x}{x^2-9}$, $\dfrac{2}{x^2+2x-15}$

16. $\dfrac{3}{2a-1}$, $\dfrac{4}{1-2a}$

17. $\dfrac{2}{a-b}$, $\dfrac{3a}{4a-4b}$

18. $\dfrac{3}{a^2-2ab+b^2}$, $\dfrac{5}{a-b}$

19. $\dfrac{7}{15a^2+34ab+15b^2}$, $\dfrac{3}{15a^2-16ab-15b^2}$

20. $\dfrac{a}{a^2-b^2}$, $\dfrac{b}{a^2+2ab+b^2}$

◆◆◆◆◆

*SOLUTIONS TO EXERCISES

I. Convert each rational expression into an equivalent rational expression with the indicated denominator or numerator.

1. $\dfrac{1}{4} = \dfrac{10}{40}$

2. $\dfrac{-2}{5} = \dfrac{12}{-30}$

3. $a = \dfrac{1}{\frac{1}{a}}$

4. $\dfrac{7}{x^2 y} = \dfrac{7xy^2}{x^3 y^3}$

5. $\dfrac{x+3}{2x+5} = \dfrac{(x+3)(x+1)}{(2x+5)(x+1)}$

6. $\dfrac{5}{9-x^2} = \dfrac{5(x+1)}{9+9x-x^2-x^3}$

7. $17 = \dfrac{17a^2}{a^2}$

8. $-3 = \dfrac{-3(2x+1)}{(2x+1)} = \dfrac{-6x-3}{2x+1}$

9. $\dfrac{x-y}{y-x} = \dfrac{y-x}{x-y}$

10. $\dfrac{7x+9}{7x^2+9x} = \dfrac{1}{\frac{7x^2+9x}{7x+9}}$

11. $\dfrac{x-4}{6} = \dfrac{x^2-16}{6(x+4)}$

12. $\dfrac{3x}{-5y^2} = \dfrac{9x}{-15y^2}$

II. Find the LCD for each pair of rational expressions, and convert each rational expression into an equivalent rational expression with the LCD as denominator.

13. $\dfrac{7}{51} = \dfrac{28}{204}$, $\dfrac{3}{68} = \dfrac{9}{204}$

Consider the denominators:
$51 = 3 \cdot 17$
$68 = 2^2 \cdot 17$
The LCD OF 51 and 68
is $2^2 \cdot 3 \cdot 17 = 204$

14. $\dfrac{3}{5ab^2c} = \dfrac{21a}{35a^2b^2c}$, $\dfrac{4}{7a^2bc} = \dfrac{20b}{35a^2b^2c}$

Consider the denominators:
The LCD of $5ab^2c$ and $7a^2bc$
is $35a^2b^2c$

15. $\dfrac{9x}{x^2-9} = \dfrac{9x(x+5)}{(x^2-9)(x+5)} = \dfrac{9x^2+45x}{x^3+5x^2-9x-45}$

$\dfrac{2}{x^2+2x-15} = \dfrac{2(x+3)}{(x^2+2x-15)(x+3)} = \dfrac{2x+6}{x^3+5x^2-9x-45}$

Consider the denominators:
$x^2 - 9 = (x + 3)(x - 3)$
$x^2 + 2x - 15 = (x + 5)(x - 3)$
The LCD of $(x^2 - 9)$ and $(x^2 + 2x -15)$
is $(x + 3)(x - 3)(x + 5)$

16. $\dfrac{3}{2a-1} = \dfrac{3}{2a-1}$, $\dfrac{4}{1-2a} = \dfrac{-4}{2a-1}$

17. $\dfrac{2}{a-b} = \dfrac{8}{4a-4b}$, $\dfrac{3a}{4a-4b}$

96

18. $\dfrac{3}{a^2-2ab+b^2} = \dfrac{3}{(a-b)^2}$, $\quad \dfrac{5}{(a-b)} = \dfrac{5(a-b)}{(a-b)^2}$

19. $\dfrac{7}{15a^2+34ab+15b^2}$, $\quad \dfrac{3}{15a^2-16ab-15b^2}$

20. $\dfrac{a}{a^2-b^2}$, $\quad \dfrac{b}{a^2+2ab+b^2}$

Consider the denominators:

$(15a^2 + 34ab + 15b^2)$
$= (3a + 5b)(5a + 3b)$

$(15a^2 -16ab - 15b^2)$
$= (3a - 5b)(5a + 3b)$

The LCD for the denominators
$= (3a + 5b)(5a + 3b)(3a - 5b)$

$\dfrac{7(3a-5b)}{(3a+5b)(5a+3b)(3a-5b)}$,

$\dfrac{3(3a+5b)}{(3a-5b)(5a+3b)(3a+5b)}$.

Consider the denominators:

$(a^2 - b^2) = (a-b)(a+b)$

$(a^2 + 2ab + b^2) = (a + b)^2$

The LCD for the denominators
$= (a - b)(a + b)^2$
$= a^3 + a^2b - ab^2 - b^3$

$\dfrac{a}{(a^2-b^2)} = \dfrac{a(a+b)}{(a-b)(a+b)^2}$,

$\dfrac{b}{(a+b)^2} = \dfrac{b(a-b)}{(a+b)^2(a-b)}$.

♦♦♦♦♦

5.4 ADDITION AND SUBTRACTION

OBJECTIVES:

*(A) To add and subtract rational numbers.
*(B) To add and subtract rational expressions.

** REVIEW

*(A) ● To add and subtract rational numbers having a common denominator, simply add or subtract the numerators.

Examples:

$a.$ $\dfrac{3}{17} + \dfrac{4}{17} = \dfrac{3+4}{17} = \dfrac{7}{17}$

$b.$ $\dfrac{7}{29} - \dfrac{5}{29} = \dfrac{7-5}{29} = \dfrac{2}{29}$

● To add or subtract rational numbers having different denominators, find the LCD and then add or subtract.
Example:

$$\dfrac{1}{12} + \dfrac{1}{30} + \dfrac{1}{24}$$

Consider the denominators:
$12 = 2 \cdot 3 \cdot 2 = 2^2 \cdot 3$
$30 = 2 \cdot 3 \cdot 5 = 2 \cdot 3 \cdot 5$
$24 = 2 \cdot 3 \cdot 2 \cdot 2 = 2^3 \cdot 3$
The LCD of 12, 30, and 24 is $2^3 \cdot 3 \cdot 5 = 120$

$$\dfrac{1}{12} + \dfrac{1}{30} + \dfrac{1}{24} = \dfrac{10}{120} + \dfrac{4}{120} + \dfrac{5}{120} = \dfrac{10+4+5}{120} = \dfrac{19}{120}$$

*(B) ● To add and subtract rational expressions having a common denominator, add or subtract the numerators.

● To add or subtract rational expressions having different denominators, find the LCD and then add or subtract.

98

Examples:

a. $\dfrac{3x}{x^2+4} + \dfrac{9x^2-3x+36}{x^2+4} = \dfrac{(3x)+(9x^2-3x+36)}{x^2+4} = \dfrac{9x^2+36}{x^2+4} = \dfrac{9(x^2+4)}{x^2+4} = 9$

b. $\dfrac{x^2-2x}{(x-1)(3x-4)} - \dfrac{x^2-2}{(x-1)(3x-4)} = \dfrac{(x^2-2x)-(x^2-2)}{(x-1)(3x-4)}$

$= \dfrac{x^2-2x-x^2+2}{(x-1)(3x-4)} = \dfrac{-2(x-1)}{(x-1)(3x-4)} = \dfrac{-2}{(3x-4)}$

◆◆◆◆◆

**EXERCISES

I. Perform the indicated operations:

1. $\dfrac{3}{15} + \dfrac{7}{15}$

2. $\dfrac{4}{9} - \dfrac{5}{9}$

3. $\dfrac{-1}{6} + \dfrac{-3}{6}$

4. $\dfrac{-5}{12} - \dfrac{8}{12}$

5. $\dfrac{11}{12} + \dfrac{2}{16}$

6. $\dfrac{-2}{3} - \dfrac{4}{15}$

7. $\dfrac{-1}{9} - \dfrac{-5}{21}$

8. $\dfrac{2}{y} + \dfrac{4}{3x}$

9. $\dfrac{5}{3x} - \dfrac{2}{9x^2}$

10. $\dfrac{3x+1}{10} - \dfrac{4-2x}{15}$

11. $\dfrac{1}{a} + \dfrac{1}{b}$

12. $\dfrac{4}{x+3} - \dfrac{1}{x+3}$

13. $\dfrac{9}{y-2} + \dfrac{2y}{y-2} + \dfrac{-13}{y-2}$

14. $1 - \dfrac{3x}{x+5}$

15. $\dfrac{x}{9} + \dfrac{21x}{6}$

16. $\dfrac{2}{2s+4} - \dfrac{3}{s^2+3s+2}$

17. $\dfrac{2}{x} - \dfrac{3}{x^2} - \dfrac{4}{x^3}$

18. $\dfrac{3a}{ab} - \dfrac{4b}{bc}$

19. $\dfrac{1}{a-2} - \dfrac{3}{a}$

20. $\dfrac{1}{xy^2} - \dfrac{1}{x^2y}$

◆◆◆◆◆

SOLUTIONS TO EXERCISES

I. Perform the indicated operations:

1. $\dfrac{3}{15}+\dfrac{7}{15}=\dfrac{3+7}{15}=\dfrac{10}{15}=\dfrac{2}{3}$

2. $\dfrac{4}{9}-\dfrac{5}{9}=\dfrac{4-5}{9}=\dfrac{-1}{9}$

3. $\dfrac{-1}{6}+\dfrac{-3}{6}=\dfrac{-1-3}{6}=\dfrac{-4}{6}=\dfrac{-2}{3}$

4. $\dfrac{-5}{12}-\dfrac{8}{12}=\dfrac{-5-8}{12}=\dfrac{-13}{12}=-1\dfrac{1}{12}$

5. $\dfrac{11}{12}+\dfrac{2}{16}=\dfrac{44}{48}+\dfrac{6}{48}=\dfrac{50}{48}=1\dfrac{2}{48}=1\dfrac{1}{24}$

6. $\dfrac{-2}{3}-\dfrac{4}{15}=\dfrac{-10-4}{15}=-\dfrac{14}{15}$

7. $\dfrac{-1}{9}-\dfrac{-5}{21}=\dfrac{-7}{63}-\dfrac{-15}{63}=\dfrac{-7-(-15)}{63}=\dfrac{8}{63}$

8. $\dfrac{2}{y}+\dfrac{4}{3x}=\dfrac{6x+4y}{3xy}$

9. $\dfrac{5}{3x}-\dfrac{2}{9x^2}=\dfrac{15x}{9x^2}-\dfrac{2}{9x^2}=\dfrac{15x-2}{9x^2}$

10. $\dfrac{3x+1}{10}-\dfrac{4-2x}{15}=\dfrac{13x-5}{30}$

11. $\dfrac{1}{a}+\dfrac{1}{b}=\dfrac{a+b}{ab}$

12. $\dfrac{4}{x+3}-\dfrac{1}{x+3}=\dfrac{3}{x+3}$

13. $\dfrac{9}{y-2}+\dfrac{2y}{y-2}+\dfrac{-13}{y-2}=\dfrac{2(y-2)}{y-2}=2$

14. $1-\dfrac{3x}{x+5}=\dfrac{x+5-3x}{x+5}=\dfrac{5-2x}{x+5}$

15. $\dfrac{x}{9}+\dfrac{21x}{6}=\dfrac{2x}{18}+\dfrac{63x}{18}=\dfrac{65x}{18}=3\dfrac{11x}{18}$

16. $\dfrac{2}{2s+4}-\dfrac{3}{s^2+3s+2}=\dfrac{s-2}{(s+1)(s+2)}$

17. $\dfrac{2}{x}-\dfrac{3}{x^2}-\dfrac{4}{x^3}=\dfrac{2x^2-3x-4}{x^3}$

18. $\dfrac{3a}{ab}-\dfrac{4b}{bc}=\dfrac{3ac-4ab}{abc}=\dfrac{3c-4b}{bc}$

19. $\dfrac{1}{a-2}-\dfrac{3}{a}=\dfrac{a-3(a-2)}{a(a-2)}=\dfrac{a-3a+6}{a(a-2)}=\dfrac{2(3-a)}{a(a-2)}$

20. $\dfrac{1}{xy^2}-\dfrac{1}{x^2y}=\dfrac{x-y}{x^2y^2}$

♦♦♦♦♦

5.5 COMPLEX FRACTIONS

OBJECTIVES:

*(A) To simplify complex fractions.

<u>REVIEW</u>

*(A) A complex fraction is a fraction that has one or more fractions in the numerator or the denominator, or both.

To simplify a complex fraction remember:

Step 1. Find the LCD of all the denominators in the complex fraction.
Step 2. Multiply the denominator and the numerator of the complex fraction by the LCD.

Examples:

a. $\dfrac{1+\dfrac{1}{x}}{1-\dfrac{1}{x}}$ *LCD of the denominators is x*

$= \dfrac{(1+\dfrac{1}{x})\cdot x}{(1-\dfrac{1}{x})\cdot x}$ *multiply the complex fraction by the LCD which is x*

$= \dfrac{x+1}{x-1}$

b. $\dfrac{\dfrac{1}{2}-\dfrac{1}{x-2}}{\dfrac{1}{x^2-4}+1}$ *LCD of 2, x-2 ,and (x²-4) is 2(x²-4)*

$= \dfrac{(\dfrac{1}{2}-\dfrac{1}{x-2})\cdot 2(x^2-4)}{(\dfrac{1}{x^2-4}+1)\cdot 2(x^2-4)}$ *multiply the complex fraction by the LCD*

$= \dfrac{(x^2-4)-2(x+2)}{2+2(x^2-4)}$

$= \dfrac{x^2-2x-8}{2(x^2-3)}$

♦♦♦♦♦

101

EXERCISES

Simplify the complex fractions.

1. $\dfrac{\dfrac{2x^2+3x-2}{2x-1}}{x+2}$

2. $\dfrac{\dfrac{x^2-9}{(x+2)^2}}{\dfrac{x+3}{x^2-4}}$

3. $\dfrac{\dfrac{x^2y-xy^2}{x+y}}{x-y}$

4. $\dfrac{\dfrac{12x^4y^3}{5z^3}}{\dfrac{4x^3y^4}{20z^5}}$

5. $\dfrac{1-\dfrac{2}{x-1}}{\dfrac{3}{x^2-1}}$

6. $\dfrac{\dfrac{3}{5}+\dfrac{9}{10}}{\dfrac{3}{4}-\dfrac{4}{3}}$

7. $\dfrac{a-3-\dfrac{3}{a}}{a+1-\dfrac{5}{a}}$

8. $\dfrac{3-\dfrac{5}{x-2}}{\dfrac{7}{x^2-4}}$

9. $\dfrac{\dfrac{a^2}{b^2}-16}{\dfrac{a}{b}-4}$

10. $\dfrac{\dfrac{x^2+x-2}{x^2+10x+21}}{\dfrac{x^2+4x+4}{x+3}}$

11. $\dfrac{\dfrac{m^2-5}{2}}{5}$

12. $\dfrac{\dfrac{1}{x-1}+\dfrac{1}{x+1}}{\dfrac{x-1}{x+1}+\dfrac{x+1}{x-1}}$

13. $1-\dfrac{x-\dfrac{3+x}{x}}{9-x^2}$

14. $\dfrac{5x+\dfrac{1}{x-1}}{3x-\dfrac{2}{1-x^2}}$

15. $\dfrac{1-\dfrac{1}{a}-\dfrac{1}{b}-\dfrac{1}{ab}}{3-\dfrac{5}{ab}}$

16. $\dfrac{\dfrac{x}{x+2}-1}{\dfrac{2x+3}{x-1}}$

17. $\dfrac{\dfrac{1}{2a-b}-\dfrac{1}{b-2a}}{\dfrac{a}{4a^2-b^2}+\dfrac{b}{2a-b}}$

18. $1-\dfrac{1}{3}-\dfrac{\dfrac{a}{a-3}}{\dfrac{a}{3-a}+1}$

19. $\dfrac{\dfrac{2}{3a+2}}{\dfrac{3}{6a^2+13a+6}}$

20. $\dfrac{\dfrac{a^2b+ba^2}{a(a+b)}}{\dfrac{a+b}{b}}$

*SOLUTIONS TO EXERCISES

Simplify the complex fractions.

1. $\dfrac{\dfrac{2x^2+3x-2}{2x-1}}{x+2} = \dfrac{(2x^2+3x-2)\cdot[x+2]}{\dfrac{2x-1}{x+2}\cdot[x+2]} = \dfrac{(2x-1)(x+2)^2}{2x-1} = (x+2)^2$

2. $\dfrac{\dfrac{x^2-9}{(x+2)^2}}{\dfrac{x+3}{x^2-4}} = \dfrac{(\dfrac{x^2-9}{(x+2)^2})\cdot[(x+2)^2(x-2)]}{(\dfrac{x+3}{x^2-4})\cdot[(x+2)^2(x-2)]} = \dfrac{(x^2-9)(x-2)}{(x+3)(x+2)} = \dfrac{(x-3)(x-2)}{x+2}$

3. $\dfrac{\dfrac{x^2y-xy^2}{x+y}}{x-y} = \dfrac{(\dfrac{x^2y-xy^2}{x+y})\cdot[x+y]}{(x-y)\cdot[x+y]} = \dfrac{xy(x-y)}{(x+y)(x-y)} = \dfrac{xy}{(x+y)}$

4. $\dfrac{\dfrac{12x^4y^3}{5z^3}}{\dfrac{4x^3y^4}{20z^5}} = \dfrac{(\dfrac{12x^4y^3}{5z^3})\cdot[20z^5]}{(\dfrac{4x^3y^4}{20z^5})\cdot[20z^5]} = \dfrac{48x^4y^3z^2}{4x^3y^4} = 12xy^{-1}z^2 = \dfrac{12xz^2}{y}$

5. $\dfrac{1-\dfrac{2}{x-1}}{\dfrac{3}{x^2-1}} = \dfrac{(1-\dfrac{2}{x-1})\cdot[x^2-1]}{\dfrac{3}{x^2-1}\cdot[x^2-1]} = \dfrac{(x^2-1)-2(x+1)}{3} = \dfrac{x^2-2x-3}{3}$

6. $\dfrac{\dfrac{3}{5}+\dfrac{9}{10}}{\dfrac{3}{4}-\dfrac{4}{3}} = \dfrac{(\dfrac{3}{5}+\dfrac{9}{10})\cdot[60]}{(\dfrac{3}{4}-\dfrac{4}{3})\cdot[60]} = \dfrac{36+54}{45-80} = \dfrac{-90}{35} = -\dfrac{18}{7}$

7. $\dfrac{a-3-\dfrac{3}{a}}{a+1-\dfrac{5}{a}} = \dfrac{((a-3)-\dfrac{3}{a})\cdot[a]}{((a+1)-\dfrac{5}{a})\cdot[a]} = \dfrac{a^2-3a-3}{a^2+a-5}$

8. $\dfrac{3-\dfrac{5}{x-2}}{\dfrac{7}{x^2-4}} = \dfrac{(3-\dfrac{5}{x-2})\cdot[x^2-4]}{(\dfrac{7}{x^2-4})\cdot[x^2-4]} = \dfrac{3(x^2-4)-5(x+2)}{7} = \dfrac{3x^2-5x-22}{7}$

9. $\dfrac{\dfrac{a^2}{b^2}-16}{\dfrac{a}{b}-4} = \dfrac{(\dfrac{a^2}{b^2}-16)\cdot[b^2]}{(\dfrac{a}{b}-4)\cdot[b^2]} = \dfrac{a^2-16b^2}{ab-4b^2} = \dfrac{(a+4b)(a-4b)}{b(a-4b)} = \dfrac{a+4b}{b}$

10. $\dfrac{\dfrac{x^2+x-2}{x^2+10x+21}}{\dfrac{x^2+4x+4}{x+3}} = \dfrac{\dfrac{(x+2)(x-1)}{(x+3)(x+7)}\cdot[(x+3)(x+7)]}{\dfrac{(x+2)^2}{(x+3)}\cdot[(x+3)(x+7)]} = \dfrac{(x+2)(x-1)}{(x+2)^2(x+7)}$

$= \dfrac{x-1}{(x+2)(x+7)}$

11. $\dfrac{\dfrac{m^2-5}{2}}{5} = \dfrac{\left(\dfrac{m^2-5}{2}\right)\cdot[2]}{5\cdot[2]} = \dfrac{m^2-5}{10}$

12. $\dfrac{\dfrac{1}{x-1}+\dfrac{1}{x+1}}{\dfrac{x-1}{x+1}+\dfrac{x+1}{x-1}} = \dfrac{\left(\dfrac{1}{x-1}+\dfrac{1}{x+1}\right)\cdot[(x+1)(x-1)]}{\left(\dfrac{x-1}{x+1}+\dfrac{x+1}{x-1}\right)\cdot[(x+1)(x-1)]} = \dfrac{2x}{2x^2+2} = \dfrac{x}{x^2+1}$

13. $1-\dfrac{x-\dfrac{3+x}{x}}{9-x^2} = 1-\dfrac{\left(x-\dfrac{3+x}{x}\right)\cdot[x]}{(9-x^2)\cdot[x]} = 1-\dfrac{x^2-(3+x)}{x(9-x^2)} = -\dfrac{(x^3+x^2-10x-3)}{x(x^2-9)}$

14. $\dfrac{5x+\dfrac{1}{x-1}}{3x-\dfrac{2}{1-x^2}} = \dfrac{\left(5x+\dfrac{1}{x-1}\right)\cdot[(x^2-1)]}{\left(3x-\dfrac{2}{1-x^2}\right)\cdot[(x^2-1)]} = \dfrac{5x^3-4x+1}{3x^3-3x+2}$

15. $\dfrac{1-\dfrac{1}{a}-\dfrac{1}{b}-\dfrac{1}{ab}}{3-\dfrac{5}{ab}} = \dfrac{\left(1-\dfrac{1}{a}-\dfrac{1}{b}-\dfrac{1}{ab}\right)\cdot[ab]}{\left(3-\dfrac{5}{ab}\right)\cdot[ab]} = \dfrac{ab-a-b-1}{3ab-5}$

16. $\dfrac{\dfrac{x}{x+2}-1}{\dfrac{2x+3}{x-1}} = \dfrac{\left(\dfrac{x}{x+2}-1\right)\cdot[(x+2)(x-1)]}{\left(\dfrac{2x+3}{x-1}\right)\cdot[(x+2)(x-1)]} = \dfrac{-2(x-1)}{2x^2+7x+6}$

17. $\dfrac{\dfrac{1}{2a-b}-\dfrac{1}{b-2a}}{\dfrac{a}{4a^2-b^2}+\dfrac{b}{2a-b}} = \dfrac{\left(\dfrac{1}{2a-b}-\dfrac{1}{b-2a}\right)\cdot[\,(2a-b)\,(2a+b)\,]}{\left(\dfrac{a}{4a^2-b^2}+\dfrac{b}{2a-b}\right)\cdot[\,(2a-b)\,(2a+b)\,]} = \dfrac{2\,(2a+b)}{a+2ab+b^2}$

18. $1-\dfrac{1}{3}-\dfrac{\left(\dfrac{a}{a-3}\right)}{\left(\dfrac{a}{3-a}+1\right)} = \dfrac{2}{3}-\dfrac{\left(\dfrac{a}{a-3}\right)\cdot[a-3]}{\left(\dfrac{-a}{a-3}+1\right)\cdot[a-3]} = \left(\dfrac{2}{3}\right)-\left(\dfrac{a}{-3}\right) = \dfrac{2+a}{3}$

19. $\dfrac{\dfrac{2}{3a+2}}{\dfrac{3}{6a^2+13a+6}} = \dfrac{\left(\dfrac{2}{3a+2}\right)\cdot[\,(2a+3)\,(3a+2)\,]}{\dfrac{3}{(3a+2)\,(2a+3)}\cdot[\,(2a+3)\,(3a+2)\,]} = \dfrac{2\,(2a+3)}{3}$

20. $\dfrac{\dfrac{a^2b+ba^2}{a\,(a+b)}}{\dfrac{a+b}{b}} = \dfrac{\dfrac{a^2b+ab^2}{a\,(a+b)}\cdot[ab\,(a+b)\,]}{\dfrac{a+b}{b}\cdot[ab\,(a+b)\,]} = \dfrac{(a^2b+ab^2)\,b}{(a+b)\cdot a\,(a+b)} = \dfrac{ab^2+b^3}{a^2+2ab+b^2} = \dfrac{b^2}{(a+b)}$

◆◆◆◆◆

5.6 SOLVING EQUATIONS

OBJECTIVES:

*(A) To multiply rational expressions by the LCD.
*(B) To find the extraneous root by checking the solution in the original equation.

** REVIEW

*(A) To solve an equation involving rational expressions, remember to multiply each side of the equation by the LCD; then simplify the equation.

Examples:

a. $\dfrac{x}{3} - \dfrac{1}{5} = \dfrac{4}{5}$ *The LCD of 3 and 5 is 15.*

$15 \cdot \left(\dfrac{x}{3}\right) - 15 \cdot \left(\dfrac{1}{5}\right) = 15 \cdot \left(\dfrac{4}{5}\right)$ *multiply each side by 15*

$5x - 3 = 12$ *simplify*
$5x = 15$
$x = 3$

If $x = 3$, *then* $\dfrac{x}{3} - \dfrac{1}{5} = \dfrac{3}{3} - \dfrac{1}{5} = 1 - \dfrac{1}{5} = \dfrac{4}{5}$.

$x = 3$ *satisfies the equation.*

b. $\dfrac{6}{x} + \dfrac{13}{x+7} = 2$ *The LCD of x and x+7 is x(x+7).*

$[x(x+7)] \cdot \dfrac{6}{x} + [x(x+7)] \cdot \dfrac{13}{x+7} = x(x+7) \cdot 2$ *multiply each side by x(x+7)*

$6(x+7) + 13x = 2x(x+7)$ *simplify*

$6x + 42 + 13x = 2x^2 + 14x$

$2x^2 - 5x - 42 = 0$

$(x - 6)(2x + 7) = 0$

$x = 6$, *or* $x = -\dfrac{7}{2}$

If $x = 6$, *then* $\dfrac{6}{x} + \dfrac{13}{x+7} = 2$ *so x=6 satisfies the equation.*

If $x = -\dfrac{7}{2}$, *then* $\dfrac{6}{x} + \dfrac{13}{x+7} = 2$ *so* $x = -\dfrac{7}{2}$ *satisfies the equation.*

*(B) When solving an equation involving rational expressions, you must check every value of x to see if it causes a 0 in the denominator. If it does, then the value of x is an extraneous root.

107

Examples:

a. $\dfrac{1}{x-2} + \dfrac{1}{x-1} = \dfrac{x-1}{x-2}$ $LCD = (x-1)(x-2)$

   $\left(\dfrac{1}{x-2} + \dfrac{1}{x-1}\right) \cdot [(x-1)(x-2)] = \dfrac{(x-1)}{(x-2)} \cdot [(x-1)(x-2)]$

   $(x-1) + (x-2) = (x-1)^2$ simplify
   $x^2 - 4x + 4 = 0$
   $(x-2)^2 = 0$
   $x = 2$
   If $x = 2$, then the denominator $(x-2)=0$

   Therefore there is no solution.

b. $\dfrac{x-3}{x} = \dfrac{-1}{x}$
   $\dfrac{x-3}{x} \cdot x = \dfrac{-1}{x} \cdot x$

   $(x-3) = -(1)$ multiply both sides by x (LCD)
   $x = 2$ $x \neq 0$

If $x = 2$, then $2(2-3)=-2$. It satisfies the equation.
$x = 2$ is the solution.

**EXERCISES

Solve each equation and check your answer.

1. $\dfrac{4}{x} = 12$ 2. $\dfrac{x}{x+1} = 6$

3. $\dfrac{3}{x} - 2 = \dfrac{1}{x}$ 4. $\dfrac{1}{x} + \dfrac{9}{4} = \dfrac{3}{x}$

5. $\dfrac{x}{2} = \dfrac{x}{3} + 1$ 6. $\dfrac{12x}{x+3} = \dfrac{3}{x+3}$

7. $\dfrac{1}{x} - x = 0$ 8. $\dfrac{x+8}{4} - \dfrac{x-2}{8} = \dfrac{x}{2}$

9. $\dfrac{3x+1}{4} - \dfrac{x}{5} = x-1$ 10. $\dfrac{x}{3} + \dfrac{3}{4} = \dfrac{x}{6} - \dfrac{7}{12}$

11. $\dfrac{3y}{5} = 2 + y$

12. $\dfrac{1}{x+2} - \dfrac{2}{x-2} = \dfrac{-11}{x^2-4}$

13. $\dfrac{7}{x-1} = \dfrac{x-1}{7}$

14. $\dfrac{3}{(x+5)(x+3)} = \dfrac{1}{(x+5)(x-3)}$

15. $\dfrac{4}{4x^2-1} + \dfrac{3}{2x+1} = \dfrac{-5}{2x-1}$

16. $\dfrac{x-3}{2} = \dfrac{2x+4}{5}$

17. $\dfrac{2}{x-9} = \dfrac{x+3}{3x-27} + 1$

18. $\dfrac{1}{x+1} - \dfrac{2}{(x+1)^2} = \dfrac{3}{x+1}$

19. $\dfrac{3x}{x} + \dfrac{x+4}{x-1} = \dfrac{-1}{x-1}$

20. $\dfrac{1}{x-11} = \dfrac{x}{3x-33} - 1$

♦♦♦♦♦

**SOLUTIONS TO EXERCISES

Solve each equation.

1. $x = 1/3$ (LCD=x)

2. $x = -(6/5)$ (LCD=(x+1))

3. $x = 1$ (LCD=x)

4. $x = (8/9)$ (LCD=4x)

5. $x = 6$ (LCD=6)

6. $x = 1/4$ (LCD=(x+3))

7. $x = \pm 1$ (LCD=x)

8. $x = 6$ (LCD=8)

9. $x = (25/9)$ (LCD=20)

10. $x = -8$ (LCD=12)

11. $y = -5$ (LCD=5)

12. $x = 5$ (LCD=x^2-4)

13. $x = 8$ or $x = -6$ (LCD=7(7(x-1))

14. $x = 6$ (LCD=(x+5)(x^2-9))

15. $x = -(3/8)$ (LCD=4x^2-1)

16. $x = 23$ (LCD=10)

17. $x = 15/2$ (LCD=3(x-9))

18. $x = -2$ (LCD=$(x+1)^2$)

19. $x = -\frac{1}{2}$ (LCD=x(x-1))

20. $x = 15$ (LCD =3(x-11))

5.7 RATIO AND PROPORTION

OBJECTIVES:

* (A) To understand ratios and proportions.

- The ratio of two numbers, a and b, with b ≠ 0, is the fraction a/b.
- A proportion is an equality with two ratios.

** REVIEW

Examples:

a. $\dfrac{a}{b} = a:b$, $b \neq 0$

b. $\dfrac{\frac{3}{5}}{\frac{4}{7}} = \dfrac{3}{5} : \dfrac{4}{7}$

c. $\dfrac{30x}{7y} = 30x:7y$

These are examples of ratios.

d. $\dfrac{a}{b} = \dfrac{c}{d} = a:b = c:d$ *where a and d are called extremes,*
                        b and c the means.

e. $\dfrac{x+y}{2} = \dfrac{3x-y}{5} = (x+y):2 = (3x-y):5$

These are examples of proportions.

- **The Extremes-Means Property:**
- **The product of the extremes is equal to the product of the means.**

a. $\dfrac{5}{4} = \dfrac{x}{2x-6}$ *apply the extremes-means property*

     $5(2x-6) = 4 \cdot x$ *5 and (2x-6) are extremes, 4 and x are means*
     $10x - 30 = 4x$
     $6x = 30$
     $x = 5$

b. $\dfrac{x}{5} = \dfrac{x+6}{3}$

     $3x = 5x+30$ *apply the extremes-means property*
     $2x = -30$
     $x = -15$

◆◆◆◆◆

EXERCISES

I. For each of the following ratios find an equivalent ratio of integers in lowest terms.

1. $\dfrac{30}{50}$

2. $\dfrac{0.15}{0.5}$

3. $\dfrac{22}{44}$

4. $\dfrac{\frac{2}{3}}{\frac{3}{2}}$

5. $\dfrac{3.5}{7}$

6. $\dfrac{5}{\frac{1}{5}}$

7. $\dfrac{7}{3.5}$

8. $\dfrac{\frac{3}{5}}{5}$

II. Solve each proportion. (Hint: apply the extremes-means property)

9. $\dfrac{3}{x}=\dfrac{x}{3}$

10. $\dfrac{-x}{4}=\dfrac{2x-7}{6}$

11. $\dfrac{10}{x}=\dfrac{2}{x+4}$

12. $\dfrac{x-2}{x+1}=\dfrac{x+2}{x-5}$

13. $\dfrac{x}{x+3}=\dfrac{x-6}{x}$

14. $\dfrac{m}{11}=\dfrac{11}{m}$

15. $\dfrac{-1}{2}=\dfrac{2}{x}$

16. $\dfrac{3}{x-3}=\dfrac{5}{x+1}$

17. $\dfrac{2a}{5}=\dfrac{4a}{10}$

18. $\dfrac{x}{x-4}=\dfrac{x}{x-5}$

19. $\dfrac{3}{2x-1}=\dfrac{2x+1}{1}$

20. $\dfrac{x-1}{x+2}=\dfrac{x+2}{x+1}$

◆◆◆◆◆

SOLUTIONS TO EXERCISES

I. For each of the following ratios find an equivalent ratio of integers in lowest terms.

1. 3/5	2. 3/10	3. 1/2	4. 4/9
5. 1/2	6. 25	7. 2	8. 3/25

II. Solve each proportion.

9. $x^2 = 9$
   $x = \pm 3$

10. $-6x = 8x - 28$
    $x = 2$

11. $10(x+4)=2x$
    $x = -5$

12. $(x-2)(x-5)=(x+1)(x+2)$
    $x^2-7x+10=x^2+3x+2$
    $10x=8, \quad x=4/5$

13. $x^2 = (x+3)(x-6)$
    $x^2 = x^2 - 3x - 18$
    $x = -6$

14. $m^2 = (11)^2$
    $m = \pm11$

15. $-x = 4$
    $x = -4$

16. $3(x+1)=5(x-3)$
    $x = 9$

17. $20a = 20a$
    a can be any real number

18. $x(x - 5)= x(x - 4)$
    $x = 0$

19. $3 = (2x + 1)(2x - 1)$
    $3 = 4x^2 - 1$
    $x = \pm 1$

20. $(x - 1)(x + 1) = (x +2)^2$
    $x^2 - 1 = x^2 + 4x + 4, \quad x = -(5/4)$

5.8 **APPLICATIONS**

OBJECTIVES:

* To solve the formula for a specific variable.

** **REVIEW**

To solve the formula for a specific variable remember to isolate the variable on one side of the equation by multipling both sides by the LCD. Then simplify.

Examples:

a. The formula $\frac{3y-5x}{5-3x} = \frac{3}{5}$, solve for y.

The extremes-means property

$5(3y-5x) = 3(5-3x)$

$15y-25x = 15-9x$

$15y = 16x+15$

$y = \frac{16x+15}{15}$

b. In the formula of example a. above, find x if y = 0

$y = \frac{16x+15}{15}$ from the result of part a

$0 = \frac{16x+15}{15}$ replace y by 0

$16x+15 = 0$

$x = \frac{-15}{16}$

◆◆◆◆◆

**<u>EXERCISES</u>

I. Solve each equation for the indicated variable.

1. $RS = PT$ for S

2. $\dfrac{1}{3m} + \dfrac{1}{5m} = \dfrac{1}{3n}$ for m

3. $\dfrac{a}{bc} = \dfrac{ef}{g}$ for f

4. $TR + TRS = I$ for T

5. $\dfrac{PV}{T} = \dfrac{T}{PV}$ for T^2

6. $\dfrac{1}{a} + \dfrac{1}{b} + \dfrac{1}{c} = 1$ for c

7. $\dfrac{x-y}{x+3} = z$ for y

8. $\dfrac{x+5}{y-3} = 4$ for x

9. $\dfrac{am-1}{2m+n} = \dfrac{1}{2}$ for n

10. $\dfrac{1}{x} + \dfrac{1}{y} = \dfrac{1}{z}$ for x

11. $\dfrac{ab}{c} = \dfrac{ef}{g}$ for g

12. $r = \dfrac{9h^2 + 5s^2}{7t}$ for t

13. $\dfrac{a}{b} = \dfrac{c}{d} = \dfrac{e}{f} = 1$ for f

14. $s = \dfrac{5}{8} rs^2 t^3$ for r

15. $hs = \dfrac{kp}{3}$ for k

16. In the formula of 10, if $y=1$ and $z=2$, find x

17. In the formula of 12, if $h=0, s=1$ and $r=1$, find t

18. In the formula of 14, if $s=1$, $t=2$, find r

19. $\dfrac{3x+y}{2} = \dfrac{2y-x}{5}$ for y

20. $A = \dfrac{B+D}{B}$ for B

◆◆◆◆◆

113

1. $RS=PT$ for S, $S=\dfrac{PT}{R}$

2. $\dfrac{1}{3m}+\dfrac{1}{5m}=\dfrac{1}{3n}$ for m, $m=\dfrac{8n}{5}$

3. $\dfrac{a}{bc}=\dfrac{ef}{g}$ for f, $f=\dfrac{ag}{bce}$

4. $TR+TRS=I$ for T, $T=\dfrac{I}{R(S+1)}$

5. $\dfrac{PV}{T}=\dfrac{T}{PV}$ for T^2, $T^2=(PV)^2$

6. $\dfrac{1}{a}+\dfrac{1}{b}+\dfrac{1}{c}=1$ for c, $c=\dfrac{ab}{ab-a-b}$

7. $\dfrac{x-y}{x+3}=z$ for y, $y=x-z(x+3)$

8. $\dfrac{x+5}{y-3}=4$ for x, $x=4y-17$

9. $\dfrac{am-1}{2m+n}=\dfrac{1}{2}$ for n, $n=2(am-m-1)$

10. $\dfrac{1}{x}+\dfrac{1}{y}=\dfrac{1}{z}$ for x, $x=\dfrac{yz}{y-z}$

11. $\dfrac{ab}{c}=\dfrac{ef}{g}$ for g, $g=\dfrac{cef}{ab}$

12. $r=\dfrac{9h^2+5s^2}{7t}$ for t, $t=\dfrac{9h^2+5s^2}{7r}$

13. $\dfrac{a}{b}-\dfrac{c}{d}-\dfrac{e}{f}=1$ for f, $f=\dfrac{bde}{ad-bc-bd}$

14. $s=\dfrac{5}{8}rs^2t^3$ for r, $r=\dfrac{8}{5st^3}$

15. $hs=\dfrac{kp}{3}$ for k, $k=\dfrac{3hs}{p}$

16. In the formula of 10,
    if $y=1$ and $z=2$, find x, $x=-2$

17. In the formula of 12,
    if $h=0$, $s=1$ and $r=1$, find t, $t=\dfrac{5}{7}$

18. In the formula of 14,
    if $s=1$, $t=2$, find r, $r=\dfrac{1}{5}$

19. $\dfrac{3x+y}{2}=\dfrac{2y-x}{5}$ for y

    $5(3x+y)=2(2y-x)$

    $15x+5y=4y-2x$

    $y=-17x$

20. $A=\dfrac{B+D}{B}$ for B, $B=\dfrac{D}{A-1}$

◆◆◆◆◆

CHAPTER V - TEST A

I. In place of the question mark, substitute an expression that will make the two rational expressions equivalent.

1. $\dfrac{1}{14} = \dfrac{?}{140}$

2. $10 = \dfrac{?}{2}$

3. $17 = \dfrac{17}{?}$

4. $\dfrac{2}{m-n} = \dfrac{?}{n-m}$

5. $\dfrac{a}{\frac{b}{5}} = \dfrac{?}{b}$

6. $\dfrac{5}{?} = \dfrac{5ab}{6cd}$

7. $\dfrac{a-1}{a+1} = \dfrac{?}{a^2-1}$

8. $\dfrac{2xy-yx}{y-x} = \dfrac{?}{x-y}$

9. $\dfrac{5a}{a^2+12a+35} = \dfrac{?}{a+5}$

II. Perform the following operations. Reduce the answers to lowest terms.

10. $\dfrac{3}{5} + \dfrac{-2}{5}$

11. $\dfrac{1}{11} - \dfrac{2}{11}$

12. $\dfrac{-3}{14} - \dfrac{9}{14}$

13. $\dfrac{1}{3} + \dfrac{2}{5}$

14. $\dfrac{3}{4} - \dfrac{1}{14}$

15. $-\dfrac{9}{8} - \dfrac{8}{7}$

16. $-\dfrac{1}{12} + \dfrac{2}{13}$

17. $\dfrac{5}{2x} + \dfrac{10}{2x}$

18. $\dfrac{3}{x-a} - \dfrac{2}{x-a}$

19. $\dfrac{3}{5x} - y$

20. $\dfrac{2}{a^3b} - \dfrac{1}{ab^3}$

♦♦♦♦♦

CHAPTER V - TEST B

I. Find the following mentally. Write down only the answer.

1. $\dfrac{3x}{5} \div 5$

2. $\dfrac{x}{y} \div \dfrac{y}{x}$

3. $\dfrac{x}{y} \div \dfrac{x}{y}$

4. $19 \div \dfrac{1}{19}$

5. $\dfrac{1}{2} \div \dfrac{1}{3} \div \dfrac{1}{4}$

6. $\dfrac{a}{b} \div \dfrac{ac}{bc}$

7. *one-fifth of* 5

8. *one-half of* $\dfrac{5}{6}$

9. *one-half of* $\dfrac{2x}{3}$

10. *two-thirds of* $\dfrac{1}{3}$

11. *one-tenth of* 5

II. Perform the indicated operations.

12. $\dfrac{3}{8} \div \dfrac{6}{4}$

13. $1 \div \dfrac{1}{2} \div \dfrac{1}{3} \div \dfrac{1}{4}$

14. $(6-y) \div \dfrac{36-y^2}{5}$

15. $\dfrac{(x^2+6x+9)}{12} \div \dfrac{(x+3)^2}{6}$

16. $\dfrac{2-5a}{5a-2} \div \dfrac{3}{20}$

17. $\dfrac{\dfrac{x+5}{3}}{15}$

18. $\dfrac{\dfrac{x^2+5x+6}{x+2}}{x+3}$

19. $\dfrac{\dfrac{m^2-n^2}{4}}{\dfrac{m+n}{m-n}}$

20. $\dfrac{5a+5b}{a} \div \dfrac{1}{3}$

♦♦♦♦♦

CHAPTER V - TEST C

I. Simplify the complex fractions.

1. $\dfrac{\dfrac{1}{3}-\dfrac{1}{4}}{\dfrac{1}{5}+\dfrac{1}{6}}$

2. $\dfrac{\dfrac{1}{10}-\dfrac{1}{5}}{\dfrac{1}{5}+\dfrac{1}{10}}$

3. $\dfrac{\dfrac{1}{2}-\dfrac{1}{3}-\dfrac{1}{4}}{\dfrac{1}{2}+\dfrac{1}{3}+\dfrac{1}{4}}$

4. $\dfrac{1-\dfrac{1}{6}}{1+\dfrac{1}{2}}$

5. $\dfrac{\dfrac{2}{3w}+\dfrac{3}{4w}}{1-\dfrac{5}{12w}}$

6. $\dfrac{\dfrac{1}{5}+\dfrac{3}{x-2}}{\dfrac{1}{x^2-4}-\dfrac{2}{x+2}}$

7. $\dfrac{\dfrac{1}{2-x}-\dfrac{2}{x-2}}{1-\dfrac{1}{2x-4}}$

8. $\dfrac{\dfrac{1}{6}-\dfrac{5}{6(2a-1)}}{\dfrac{1}{6(2a-1)}}$

9. $\dfrac{m^2-\dfrac{5}{(m-1)}}{1-\dfrac{4}{m^2-1}}$

II. Solve each equation.

10. $\dfrac{m}{5}=\dfrac{7}{15}$

11. $\dfrac{2}{3x-9}-\dfrac{3}{x-3}=\dfrac{7}{6}$

12. $\dfrac{3}{x-5}=\dfrac{-9}{x+5}$

13. $3+\dfrac{2}{2x+1}=\dfrac{3}{2x+1}$

14. $\dfrac{1}{x}+\dfrac{1}{2}+\dfrac{1}{3}=\dfrac{13}{12(x-3)}$

15. $\dfrac{x-1}{3}\div\dfrac{x^2-1}{5}=\dfrac{5}{x+7}$

16. $\dfrac{x}{1-x}+\dfrac{x-1}{x}=\dfrac{-5}{6}$

17. $\dfrac{2x}{3}=\dfrac{3}{2x}$

18. $\dfrac{2}{2x+3}-\dfrac{3}{2x-3}=\dfrac{5}{3}$

19. $\dfrac{1+\dfrac{x}{2}+\dfrac{x}{3}}{1-\dfrac{x}{2}-\dfrac{x}{3}}=1$

20. $\dfrac{x+3}{x-3}=\dfrac{3}{5}$

♦♦♦♦♦

CHAPTER V - TEST A - SOLUTIONS

I. In place of the question mark, substitute an expression that will make these rational expressions equivalent.

1. $\dfrac{1}{14} = \dfrac{1 \cdot 10}{14 \cdot 10} = \dfrac{10}{140}$

2. $10 = \dfrac{10 \cdot 2}{1 \cdot 2} = \dfrac{20}{2}$

3. $17 = \dfrac{17}{1}$

4. $\dfrac{2}{m-n} = \dfrac{-2}{n-m}$

5. $\dfrac{a}{\frac{b}{5}} = a \cdot \dfrac{5}{b} = \dfrac{5a}{b}$

6. $\dfrac{5}{\frac{6cd}{ab}} = \dfrac{5ab}{6cd}$

7. $\dfrac{a-1}{a+1} = \dfrac{(a-1)(a-1)}{(a+1)(a-1)} = \dfrac{(a-1)^2}{a^2-1}$

8. $\dfrac{2xy-yx}{y-x} = \dfrac{-xy}{x-y}$

9. $\dfrac{5a}{a^2+12a+35} = \dfrac{5a}{(a+5)(a+7)}$

$= \dfrac{\frac{5a}{(a+7)}}{a+5}$

II. Perform the following operations. Reduce the answers to lowest terms.

10. $\dfrac{3}{5} + \dfrac{-2}{5} = \dfrac{3-2}{5} = \dfrac{1}{5}$

11. $\dfrac{1}{11} - \dfrac{2}{11} = \dfrac{1-2}{11} = \dfrac{-1}{11}$

12. $\dfrac{-3}{14} - \dfrac{9}{14} = \dfrac{-3-9}{14} = \dfrac{-6}{7}$

13. $\dfrac{1}{3} + \dfrac{2}{5} = \dfrac{5+6}{15} = \dfrac{11}{15}$

14. $\dfrac{3}{4} - \dfrac{1}{14} = \dfrac{21-2}{28} = \dfrac{19}{28}$

15. $-\dfrac{9}{8} - \dfrac{8}{7} = \dfrac{-63-64}{56} = -\dfrac{127}{56}$

16. $-\dfrac{1}{12} + \dfrac{2}{13} = \dfrac{-13+24}{156} = \dfrac{11}{156}$

17. $\dfrac{5}{2x} + \dfrac{10}{2x} = \dfrac{5+10}{2x} = \dfrac{15}{2x}$

18. $\dfrac{3}{x-a} - \dfrac{2}{x-a} = \dfrac{1}{x-a}$

19. $\dfrac{3}{5x} - y = \dfrac{3-y(5x)}{5x} = \dfrac{3-5xy}{5x}$

20. $\dfrac{2}{a^3b} - \dfrac{1}{ab^3} = \dfrac{2b^2-a^2}{a^3b^3}$

♦♦♦♦♦

I. Find the following mentally. Write down only the answer.

1. $\dfrac{3x}{5} \div 5 = \dfrac{3x}{5} \cdot \dfrac{1}{5} = \dfrac{3x}{25}$

2. $\dfrac{x}{y} \div \dfrac{y}{x} = \dfrac{x}{y} \cdot \dfrac{x}{y} = \dfrac{x^2}{y^2}$

3. $\dfrac{x}{y} \div \dfrac{x}{y} = \dfrac{x}{y} \cdot \dfrac{y}{x} = 1$

4. $19 \div \dfrac{1}{19} = 19 \cdot 19 = 361$

5. $\dfrac{1}{2} \div \dfrac{1}{3} \div \dfrac{1}{4} = \dfrac{1}{2} \cdot 3 \cdot 4 = 6$

6. $\dfrac{a}{b} \div \dfrac{ac}{bc} = \dfrac{a}{b} \cdot \dfrac{bc}{ac} = 1$

7. *one-fifth of 5 is* $\dfrac{1}{5} \cdot 5 = 1$

8. *one-half of* $\dfrac{5}{6}$ *is* $\dfrac{1}{2} \cdot \dfrac{5}{6} = \dfrac{5}{12}$

9. *one-half of* $\dfrac{2x}{3}$ *is* $\dfrac{1}{2} \cdot \dfrac{2x}{3} = \dfrac{x}{3}$

10. *two-thirds of* $\dfrac{1}{3}$ *is* $\dfrac{2}{3} \cdot \dfrac{1}{3} = \dfrac{2}{9}$

11. *one-tenth of 5 is* $\dfrac{1}{10} \cdot 5 = \dfrac{1}{2}$

II. Perform the indicated operations

12. $\dfrac{3}{8} \div \dfrac{6}{4} = \dfrac{3}{8} \cdot \dfrac{4}{6} = \dfrac{1}{4}$

13. $1 \div \dfrac{1}{2} \div \dfrac{1}{3} \div \dfrac{1}{4} = 1 \cdot 2 \cdot 3 \cdot 4 = 24$

14. $(6-y) \div \dfrac{36-y^2}{5} = (6-y) \cdot \dfrac{5}{36-y^2} = \dfrac{5}{6+y}$

15. $\dfrac{(x^2+6x+9)}{12} \div \dfrac{(x+3)^2}{6}$
$= \dfrac{(x+3)^2}{12} \cdot \dfrac{6}{(x+3)^2} = \dfrac{1}{2}$

16. $\dfrac{2-5a}{5a-2} \div \dfrac{3}{20} = \dfrac{-(5a-2)}{5a-2} \cdot \dfrac{20}{3} = -\dfrac{20}{3} = -6\dfrac{2}{3}$

17. $\dfrac{\frac{x+5}{3}}{15} = \dfrac{x+5}{3} \cdot \dfrac{1}{15} = \dfrac{x+5}{45}$

18. $\dfrac{x^2+5x+6}{\frac{x+2}{x+3}} = (x+2)(x+3) \cdot \dfrac{x+3}{x+2} = (x+3)^2$

19. $\dfrac{\frac{m^2-n^2}{4}}{\frac{m+n}{m-n}} = \dfrac{(m-n)(m+n)}{4} \cdot \dfrac{m-n}{m+n}$
$= \dfrac{(m-n)^2}{4}$

20. $\dfrac{5a+5b}{a} \div \dfrac{1}{3} = \dfrac{5(a+b)}{a} \cdot 3 = \dfrac{15(a+b)}{a}$

♦♦♦♦♦

CHAPTER V - TEST C - SOLUTIONS

I. Simplify the complex fractions.

1. $\dfrac{\frac{1}{3}-\frac{1}{4}}{\frac{1}{5}+\frac{1}{6}} = \dfrac{(\frac{1}{3}-\frac{1}{4})\cdot[60]}{(\frac{1}{5}+\frac{1}{6})\cdot[60]} = \dfrac{20-15}{12+10} = \dfrac{5}{22}$

2. $\dfrac{\frac{1}{10}-\frac{1}{5}}{\frac{1}{5}+\frac{1}{10}} = \dfrac{(\frac{1}{10}-\frac{1}{5})\cdot[10]}{[\frac{1}{5}+\frac{1}{10}]\cdot[10]} = \dfrac{1-2}{1+2} = -\dfrac{1}{3}$

3. $\dfrac{\frac{1}{2}-\frac{1}{3}-\frac{1}{4}}{\frac{1}{2}+\frac{1}{3}+\frac{1}{4}} = \dfrac{(\frac{1}{2}-\frac{1}{3}-\frac{1}{4})\cdot[12]}{(\frac{1}{2}+\frac{1}{3}+\frac{1}{4})\cdot[12]} = \dfrac{6-4-3}{6+4+3} = -\dfrac{1}{13}$

4. $\dfrac{1-\frac{1}{6}}{1+\frac{1}{2}} = \dfrac{(1-\frac{1}{6})\cdot[6]}{(1+\frac{1}{2})\cdot[6]} = \dfrac{6-1}{6+3} = \dfrac{5}{9}$

5. $\dfrac{\frac{2}{3w}+\frac{3}{4w}}{1-\frac{5}{12w}} = \dfrac{(\frac{2}{3w}+\frac{3}{4w})\cdot[12w]}{(1-\frac{5}{12w})\cdot[12w]} = \dfrac{8+9}{12w-5} = \dfrac{17}{12w-5}$

6. $\dfrac{\frac{1}{5}+\frac{3}{x-2}}{\frac{1}{x^2-4}-\frac{2}{x+2}} = \dfrac{[5(x^2-4)]\cdot(\frac{1}{5}+\frac{3}{x-2})}{[5(x^2-4)]\cdot(\frac{1}{x^2-4}-\frac{2}{x+2})} = \dfrac{(x^2-4)+15(x+2)}{5-10(x-2)} = \dfrac{x^2+15x+26}{-10x+25}$

7. $\dfrac{\frac{1}{2-x}-\frac{2}{x-2}}{1-\frac{1}{2x-4}} = \dfrac{(\frac{-1}{x-2}-\frac{2}{x-2})\cdot[2(x-2)]}{(1-\frac{1}{2(x-2)})\cdot[2(x-2)]} = \dfrac{-2-4}{2(x-2)-1} = \dfrac{-6}{2x-5}$

8. $\dfrac{\frac{1}{6}-\frac{5}{6(2a-1)}}{\frac{1}{6(2a-1)}} = \dfrac{(2a-1)-5}{1} = 2a-6$

9. $\dfrac{m^2-\frac{5}{(m-1)}}{1-\frac{4}{m^2-1}} = \dfrac{(m^2-\frac{5}{(m-1)})\cdot[m^2-1]}{(1-\frac{4}{m^2-1})\cdot[m^2-1]} = \dfrac{m^2(m^2-1)-5(m+1)}{(m^2-1)-4} = \dfrac{(m+1)(m^3-m^2-5)}{m^2-5}$

♦♦♦♦♦

II. Solve each equation.

10.

$$\frac{m}{5} = \frac{7}{15} \quad (LCD=15)$$

$$15 \cdot \frac{m}{5} = 15 \cdot \frac{7}{15}$$

$$3m=7$$

$$m=\frac{7}{3}$$

11.

$$\frac{2}{3x-9} - \frac{3}{x-3} = \frac{7}{6} \quad (LCD=6(x-3))$$

$$[6(x-3)] \cdot \left(\frac{2}{3(x-3)} - \frac{3}{(x-3)}\right) = 6(x-3) \cdot \frac{7}{6}$$

$$4-18=7(x-3)$$

$$7x=7$$

$$x=1$$

12.

$$\frac{3}{x-5} = \frac{-9}{x+5} \quad (LCD=(x^2-25))$$

$$[(x^2-25)] \cdot \frac{3}{(x-5)} = [(x^2-25)] \cdot \frac{-9}{x+5}$$

$$3(x+5)=-9(x-5)$$

$$12x=45-15=30$$

$$x=\frac{5}{2}$$

13.

$$3+\frac{2}{2x+1} = \frac{3}{2x+1} \quad (LCD=(2x+1))$$

$$[2x+1] \cdot \left(3+\frac{2}{2x+1}\right) = [2x+1] \cdot \frac{3}{2x+1}$$

$$3(2x+1)+2=3$$

$$6x=-2$$

$$x=-\frac{1}{3}$$

14.

$$\frac{1}{x} + \frac{1}{2} + \frac{1}{3} = \frac{13}{12(x-3)}$$

$$[12x(x-3)] \cdot \left(\frac{1}{x}+\frac{5}{6}\right) = [12x(x-3)] \cdot \frac{13}{12(x-3)}$$

$$12(x-3)+10x(x-3)=13x$$

$$12x-36+10x^2-30x=13x$$

$$10x^2-31x-36=0$$

$$(x-4)(10x+9)=0$$

$$x=4, \quad x=-\frac{9}{10}$$

15.

$$\frac{x-1}{3} \div \frac{x^2-1}{5} = \frac{5}{x+7}$$

$$\left(\frac{x-1}{3}\right) \cdot \left(\frac{5}{(x-1)(x+1)}\right) = \frac{5}{x+7}$$

$$\frac{5}{3(x+1)} = \frac{5}{x+7}$$

$$(x+7)=3(x+1)$$

$$x+7=3x+3$$

$$2x=4$$

$$x=2$$

16.

$$\frac{x}{1-x} + \frac{x-1}{x} = \frac{-5}{6} \qquad (LCD = 6x(1-x))$$

$$[6x(1-x)] \cdot \left(\frac{x}{1-x} + \frac{x-1}{x}\right) = [6x(1-x)] \cdot \left(\frac{-5}{6}\right)$$

$$6x^2 - 6(1-x)^2 = -5x(1-x)$$

$$6x^2 - 6 + 12x - 6x^2 = -5x + 5x^2$$

$$5x^2 - 17x + 6 = 0$$

$$(x-3)(5x-2) = 0$$

$$x = 3, \quad x = \frac{2}{5}$$

17.

$$\frac{2x}{3} = \frac{3}{2x}$$

$$4x^2 = 9$$

$$x = \pm\frac{3}{2}$$

18.

$$\frac{2}{2x+3} - \frac{3}{2x-3} = \frac{5}{3} \qquad (LCD = 3(4x^2-9))$$

$$[3(4x^2-9)] \cdot \left(\frac{2}{2x+3} - \frac{3}{2x-3}\right) = [3(4x^2-9)] \cdot \frac{5}{3}$$

$$6(2x-3) - 9(2x+3) = 5(4x^2-9)$$

$$12x - 18 - 18x - 27 = 20x^2 - 45$$

$$20x^2 + 6x = 0$$

$$x(10x+3) = 0$$

$$x = 0, \quad x = -\frac{3}{10}$$

19.

$$\frac{1 + \frac{x}{2} + \frac{x}{3}}{1 - \frac{x}{2} - \frac{x}{3}} = 1 \qquad (LCD = 6)$$

$$\frac{6 \cdot \left(1 + \frac{x}{2} + \frac{x}{3}\right)}{6 \cdot \left(1 - \frac{x}{2} - \frac{x}{3}\right)} = 1$$

$$\frac{6 + 3x + 2x}{6 - 3x - 2x} = 1$$

$$6 + 5x = 6 - 5x$$

$$10x = 0$$

$$x = 0$$

20.

$$\frac{x+3}{x-3} = \frac{3}{5}$$

$$5(x+3) = 3(x-3)$$

$$5x + 15 = 3x - 9$$

$$2x = -24$$

$$x = -12$$

CHAPTER 6

POWERS AND ROOTS

6.1 <u>POSITIVE INTEGRAL EXPONENTS</u>

<u>OBJECTIVES</u>:

*(A) To review exponents.
*(B) To raise an exponential expression to a power.
*(C) To understand the power of a product and the power of a quotient.

** REVIEW

*(A) $a^n = a \cdot a \cdot a \cdot a \cdots a$

a^n represents **a** multiplied by itself **n** times.
a^n is called an exponential expression.
a is called the base and n is the exponent or power.

Examples:

    a. $a^6 = a \cdot a \cdot a \cdot a \cdot a \cdot a$
    b. $x^4 = x \cdot x \cdot x \cdot x$
    c. $(-5)^2 = (-5)(-5) = 25$

● **Product rule:** If m and n are integers, then

$$a^m \cdot a^n = a^{(m+n)}$$

● **Quotient rule:** If n is a positive integer and $a \neq 0$, then

$$a^{-n} = \frac{1}{a^n}$$

Therefore $\dfrac{a^m}{a^n} = a^{m-n}$ if $m \geq n$

$\dfrac{a^m}{a^n} = \dfrac{1}{a^{n-m}}$ if $m < n$

● **Zero rule:** we define $a^0 = 1$ if $a \neq 0$
  Examples:

    a. $3^2 \cdot 3^5 = 3^{2+5} = 3^7$ *product rule*

    b. $\dfrac{m^6}{m^2} = m^{6-2} = m^4$ *quotient rule*

$c.$ $\dfrac{a^2}{a^5} = a^2 \cdot a^{-5} = a^{2-5} = \dfrac{1}{a^3}$

$d.$ $\dfrac{(7^3 \cdot 7^{-2})}{7^3} \cdot 7^0 = \dfrac{7^{3-2}}{7^3} \cdot 7^0 = \dfrac{7}{7} \cdot \dfrac{1}{7^2} \cdot 1 = \dfrac{1}{7^2} = \dfrac{1}{49}$

$e.$ $\dfrac{(7ab^3) \cdot a^5 b}{3a^4 b^2 a^2} = \dfrac{7a^{1+5}b^{3+1}}{3a^{4+2}b^2} = \dfrac{7a^6 b^4}{3a^6 b^2} = \dfrac{7}{3}a^{6-6}b^{4-2} = \dfrac{7}{3}b^2$

*(B) To raise an exponential expression to a power, apply the power rule. That is,
If m and n are nonnegative integers, and a≠0, then

$$(a^m)^n = a^{(m \cdot n)} = a^{mn}$$

Examples:

Simplify the following:

$a.$ $2a^5 (a^2)^3 = 2a^5 (a^{2 \cdot 3}) = 2a^5 \cdot a^6 = 2a^{11}$

$b.$ $\dfrac{(5^2)^4 \cdot 5^3}{5^2 \cdot 5^{10}} = \dfrac{5^8 \cdot 5^3}{5^2 \cdot 5^{10}} = \dfrac{5^{8+3}}{5^{2+10}} = \dfrac{5^{11}}{5^{12}} = a^{11-12} = a^{-1} = \dfrac{1}{a}$

$c.$ $\dfrac{2(x^4)^3}{8x^5} = \dfrac{x^{12}}{4x^5} = \dfrac{x^{12-5}}{4} = \dfrac{x^7}{4}$

*(C) To raise a product to a power, apply the power of a product rule.

● Power of a product rule.
If a and b are real numbers and n is a positive integer, then
$$(ab)^n = a^n b^n$$

Examples:

a. $(3xy^2)^3$ b. $(-7a^2 b^5 c)^3$ c. $(a^2 a^{-3})^4$

Solutions:

a. $(3xy^2)^3 = 3^3 x^3 (y^2)^3 = 27x^3 y^6$

b. $(-7a^2 b^5 c)^3 = (-7)^3 (a^2)^3 (b^5)^3 (c)^3 = -343a^6 b^{15} c^3$

c. $(a^2 a^{-3})^4 = (a^{2-3})^4 = a^{-4} = 1/(a^4)$

124

*(D) To raise a quotient to a power, apply the power of a quotient rule.

If a and b are real numbers, b≠0 and n is a positive integer, then

$$\left(\frac{a}{b}\right)^n = \frac{a^n}{b^n}$$

Examples:

a. $\left(\dfrac{3}{(4a^4)}\right)^2 = \dfrac{3^2}{(4a^4)^2} = \dfrac{9}{16a^8}$

b. $\left(\dfrac{m^2}{2n^5}\right)^3 = \dfrac{(m^2)^3}{(2n^5)^3} = \dfrac{m^6}{8m^{15}} = \dfrac{1}{8m^9}$

♦♦♦♦♦

**EXERCISES

Simplify the exponential expressions.

1. $(-2)^0$
2. 5^2
3. $(-2)^2(-2)^3$

4. 7^3
5. $2^2 \cdot 2^{-5}$
6. $a^{-5} \cdot a^5$

7. $(8-6)^2$
8. $(-3)^3 \cdot (-3)^2$
9. $(2t^2)^3$

10. $(-a^2b^5)^3$
11. $-2a^3b^2a^5b^7$
12. $(-2a^2b^3)^4$

13. $\left(\dfrac{a^2}{b^3}\right)^2$
14. $\left(\dfrac{2x^3}{5y^2}\right)^2$
15. $\dfrac{2x^2y^5}{x^5y^4}$

16. $\left(\dfrac{mn^4}{t^2}\right)^3$
17. $(-y^2)^3$
18. $(-3a^2)^{-5}$

19. $\left(\dfrac{3m^2 \cdot 6m^3}{9m^2}\right)^3$
20. $\dfrac{(ab^2)^3(a^2b^3)^2}{(a^2b)^2}$

**SOLUTIONS TO EXERCISES

1. $(-2)^0 = 1$
2. $5^2 = 25$

3. $(-2)^2(-2)^3 = (4)(-8) = -32$
4. $7^3 = 343$

5. $2^2 \cdot 2^{-5} = 2^{2-5} = 1/8$
6. $a^{-5} \cdot a^5 = a^0 = 1$

7. $(8-6)^2 = 2^2 = 4$
8. $(-3)^3 \cdot (-3)^2 = (-3)^5 = -243$

125

9. $(2t^2)^3 = 2^3 (t^2)^3 = 8t^6$

10. $(-a^2b^5)^3 = -a^6b^{15}$

11. $-2a^3b^2a^5b^7 = -2a^{3+5}b^{2+7} = -2a^8b^9$

12. $(-2a^2b^3)^4 = (-2)^4(a^2)^4(b^3)^4 = 16a^8b^{12}$

13. $\left(\dfrac{a^2}{b^3}\right)^2 = \dfrac{(a^2)^2}{(b^3)^2} = \dfrac{a^4}{b^6}$

14. $\left(\dfrac{2x^3}{5y^2}\right)^2 = \dfrac{4x^6}{25y^4}$

15. $\dfrac{2x^2y^5}{9x^5y^4} = \dfrac{2y^{5-4}}{9x^{5-2}} = \dfrac{2y}{9x^3}$

16. $\left(\dfrac{mn^4}{t^2}\right)^3 = \dfrac{m^3n^{12}}{t^6}$

17. $(-y^2)^3 = -y^6$

18. $(-3a^2)^{-5} = (-3)^{-5}(a^2)^{-5}$

$$= -243a^{-10} = \dfrac{-243}{a^{10}}$$

19. $\left(\dfrac{3m^2 \cdot 6m^3}{9m^2}\right)^3 = \left(\dfrac{18m^2m^3}{9m^2}\right)^3$

20. $\dfrac{(ab^2)^3(a^2b^3)^2}{(a^2b)^2} = \dfrac{a^3b^6a^4b^6}{a^4b^2}$

$= (2m^3)^3 = 8m^9$

$= a^{7-4}b^{12-2} = a^3b^{10}$

◆◆◆◆◆

6.2 **ROOTS**

OBJECTIVES:

*(A) To understand the basic concepts of exponents.
*(B) To understand the product rule for radicals.
*(C) To understand the quotient rule for radicals.

** **REVIEW**

*(A) • If a and b are real numbers such that $a^n = b$, then a is an nth root of b where n > 0.
   • If b is positive, then there is only one positive number a such that $a^n = b$. $\sqrt[n]{b}$ is called the principal nth root of b.

Examples:

(a). $5^2 = 25$ 5 is a square root of 25

(b). $(-5)^2 = 25$ - 5 is a square root of 25

(c). $(2)^4 = 16$ 2 is a fourth root of 16

(d). $(-2)^4 = 16$ -2 is a fourth root of 16

(e). $(3)^3 = 27$ 3 is a cube root of 27

(f). $(-5)^3 = -125$ -5 is a cube root of -125

   • There are **two even roots of any positive number.**
   See examples (a) (b) (c) and (d).

   • There is **only one real odd root of any real number.**
   See examples (e) and (f).

   The principal square root of 25 is 5 but not -5.
   The principal fourth root of 16 is 2 but not -2.
   The principal cube root of 27 is 3.

   • The symbol $\sqrt[n]{a}$ is used to signify the nth root and represents only the positive root of the real number a.

Examples:

a. $\sqrt[n]{a}$ means to find a real number which, when raised to the nth power, gives a.

   • If n is a positive **even** integer and a is positive, then $\sqrt[n]{a}$ is the principal nth root of a.

   • If n is a positive **odd** integer, then $\sqrt[n]{a}$ is the nth root of a.

   • $\sqrt[n]{0} = 0$.

127

b. $\sqrt[2]{81} = \sqrt{81} = 9$ where 9 is the principal square root of 81

c. $\sqrt[5]{(-32)} = -2$ where -2 is the fifth root of -32

Consider the expressions below:

x^2, x^4, x^6, x^8 ... These are perfect squares

x^3, x^6, x^9, x^{12} ... These are perfect cubes

a. $\sqrt{x^2} = x,$ $\sqrt{x^8} = \sqrt{(x^4)^2} = x^4,$ $\sqrt{x^{16}} = \sqrt{(x^8)^2} = x^8$

b. $\sqrt[3]{x^3} = x,$ $\sqrt[3]{x^6} = \sqrt[3]{(x^3)^2} = x^2,$ $\sqrt[3]{x^{27}} = \sqrt[3]{(x^9)^3} = x^9$

*(B) *The product rule for radicals.*

$$\sqrt[n]{ab} = \sqrt[n]{a}\,\sqrt[n]{b}, \text{ for any real numbers } a, b \text{ and } n.$$

*(C) *The quotient rule for radicals.*

$$\sqrt[n]{\frac{a}{b}} = \frac{\sqrt[n]{a}}{\sqrt[n]{b}}, \text{ for any real numbers } a, b \text{ and } n \text{ where } b \neq 0.$$

Examples:

a. $\sqrt{3xy} = \sqrt{3}\,\sqrt{x}\cdot\sqrt{y}$

b. $\sqrt{16x^4y^2} = \sqrt{16}\sqrt{x^4}\sqrt{y^2} = 4x^2y$

c. $\sqrt[3]{\dfrac{m}{8n}} = \dfrac{\sqrt[3]{m}}{\sqrt[3]{8n}} = \dfrac{\sqrt[3]{m}}{2\sqrt[3]{n}}$

d. $\sqrt[5]{\dfrac{x^{15}}{x^{20}}} = \dfrac{\sqrt[5]{x^{15}}}{\sqrt[5]{x^{20}}} = \dfrac{\sqrt[5]{(x^3)^5}}{\sqrt[5]{(x^4)^5}} = \dfrac{x^3}{x^4} = \dfrac{1}{x}$

♦♦♦♦♦

**EXERCISES

I. Find the following roots or powers.

1. $(-2)^3$ 2. 3^3 3. 4^2

4. $\sqrt[5]{-32}$ 5. $-\sqrt[10]{1}$ 6. $\sqrt[5]{-1}$

7. $\sqrt[6]{729}$ 8. $\sqrt[5]{y^{10}}$ 9. $\sqrt[3]{a^3b^6c^9}$

10. $-\sqrt[4]{81x^{16}}$ 11. $-\sqrt{121}$ 12. $\sqrt[4]{16a^8b^{12}}$

II. Use the product rule and/or quotient rule for radicals to simplify each expression.

13. $\sqrt{16y^3}$ 14. $\sqrt{25a^4b^2}$ 15. $\sqrt[3]{\dfrac{-8y^5}{81x^6}}$

16. $\sqrt{\dfrac{a}{16}}$ 17. $\sqrt[4]{\dfrac{a^{16}}{16b^4}}$ 18. $\sqrt[3]{\dfrac{-729}{a^3b^6}}$

19. $-\sqrt[3]{1000k^3}$ 20. $\sqrt[5]{\dfrac{a^{15}}{-32b^{30}}}$

♦♦♦♦♦

SOLUTIONS TO EXERCISES

I. Find the following roots or powers.

1. $(-2)^3 = -8$ 2. $3^3 = 27$ 3. $4^2 = 16$

4. $\sqrt[5]{-32} = -2$ 5. $-\sqrt[10]{1} = -1$ 6. $\sqrt[5]{-1} = -1$

7. $\sqrt[6]{729} = 3$ 8. $\sqrt[5]{y^{10}} = y^{\frac{10}{5}} = y^2$ 9. $\sqrt[3]{a^3b^6c^9} = ab^2c^3$

10. $-\sqrt[4]{81x^{16}} = -3x^4$ 11. $-\sqrt{121} = -11$ 12. $\sqrt[4]{16a^8b^{12}} = 2a^2b^3$

II. Use the product rule and/or quotient rule for radicals to simplify each expression.

13. $\sqrt{16y^3} = 4y\sqrt{y}$

14. $\sqrt{25a^4b^2} = \sqrt{(5a^2b)^2} = 5a^2b$

15. $\sqrt[3]{\dfrac{-8y^5}{81x^6}} = \dfrac{\sqrt[3]{(-2y)^3y^2}}{\sqrt[3]{(3x^2)^33}} = \dfrac{-2y\sqrt[3]{y^2}}{3x^2\sqrt[3]{3}}$

16. $\sqrt{\dfrac{a}{16}} = \dfrac{\sqrt{a}}{\sqrt{16}} = \dfrac{\sqrt{a}}{4}$

17. $\sqrt[4]{\dfrac{a^{16}}{16b^4}} = \dfrac{\sqrt[4]{(a^4)^4}}{\sqrt[4]{(2b)^4}} = \dfrac{a^4}{2b}$

18. $\sqrt[3]{\dfrac{-729}{a^3b^6}} = \dfrac{\sqrt[3]{(-9)^3}}{\sqrt[3]{(ab^2)^3}} = \dfrac{-9}{ab^2}$

19. $-\sqrt[3]{1000k^3} = -\sqrt[3]{(10k)^3} = -10k$

20. $\sqrt[5]{\dfrac{a^{15}}{-32b^{30}}} = \sqrt[5]{\dfrac{(a^3)^5}{(-2b^6)^5}} = \dfrac{a^3}{-2b^6}$

◆◆◆◆◆

130

6.3 SIMPLIFYING SQUARE ROOTS

OBJECTIVES:

*(A) To rationalize denominators.
*(B) To simplify square roots involving variables.

** REVIEW

*(A) To rationalize the denominator, multiply the numerator and the denominator of the radicand by the number that will make the denominator a perfect cube.

Example:

$$\sqrt[3]{\frac{2}{7}} = \sqrt[3]{\frac{2}{7} \cdot \left(\frac{7^2}{7^2}\right)} = \sqrt[3]{\frac{2 \cdot 7^2}{7^3}} = \frac{\sqrt[3]{98}}{7}$$

the radicand is $\frac{2}{7}$ and$\left(\frac{7^2}{7^2}\right)$ is the number that will make the

denominator a perfect n^{th} power

*(B) To simplify square roots involving variables, see if there is a perfect square in the expression.

Examples:

a. $\sqrt{x^7} = \sqrt{x^6 \cdot x} = \sqrt{(x^3)^2 \cdot x} = x^3\sqrt{x}$

b. $\sqrt{12x^6} = \sqrt{4x^6 \cdot 3} = \sqrt{(2x^3)^2 \cdot 3} = 2x^3\sqrt{3}$

c. $\sqrt{72m^5n^9} = \sqrt{6^2 \cdot 2(m^2)^2 \cdot m \cdot (n^4)^2 \cdot n} = \sqrt{(6m^2n^4)^2 \cdot 2mn} = 6m^2n^4\sqrt{2mn}$

**EXERCISES

Simplify the following:

1. $\sqrt{300}$ 2. $\sqrt{242}$ 3. $\sqrt{12a^5b^6}$ 4. $\sqrt{\frac{3}{4}}$

5. $\frac{9}{12}$ 6. $\frac{\sqrt{6x^2}}{2y}$ 7. $\sqrt{54a^2b^3}$ 8. $\frac{2}{\sqrt{3}}$

9. $\sqrt{\frac{3}{17}}$ 10. $\sqrt[3]{\frac{3}{5}}$ 11. $\frac{2}{\sqrt[3]{5}}$ 12. $\frac{3x\sqrt{y}}{\sqrt{2x}}$

131

$13. \sqrt{\dfrac{9x^3}{5x}}$ $\qquad 14. \dfrac{a}{\sqrt{a^3}}$ $\qquad 15. \sqrt{14ab^3}$ $\qquad 16. \sqrt{24xy^5}$

$17. \sqrt{\dfrac{15b}{a}}$ $\qquad 18. \sqrt{200xy^5}$ $\qquad 19. \sqrt{8x^3y^5z^2}$ $\qquad 20. \dfrac{10}{\sqrt{2t}}$

**SOLUTIONS TO EXERCISES

Simplify the following:

$1. \sqrt{300} = \sqrt{100\cdot 3} = \sqrt{(10)^2\cdot 3} = 10\sqrt{3}$ $\qquad 2. \sqrt{242} = \sqrt{121\cdot 2} = \sqrt{(11)^2\cdot 2} = 11\sqrt{2}$

$3. \sqrt{12^5b^6} = \sqrt{(2a^2b^3)^2\cdot(3a)} = 2a^2b^3\sqrt{3a}$ $\qquad 4. \sqrt{\dfrac{3}{4}} = \dfrac{\sqrt{3}}{\sqrt{4}} = \dfrac{\sqrt{3}}{2}$

$5. \sqrt{\dfrac{9}{12}} = \dfrac{\sqrt{3^2}}{\sqrt{(2)^2\cdot 3}} = \left(\dfrac{3}{2\sqrt{3}}\right)\cdot\left(\dfrac{\sqrt{3}}{\sqrt{3}}\right) = \dfrac{\sqrt{3}}{2}$ $\qquad 6. \sqrt{\dfrac{6x^2}{2y}} = \sqrt{\dfrac{6x^2}{2y}\cdot\left(\dfrac{2y}{2y}\right)} = \dfrac{2x\sqrt{3y}}{2y}$

$7. \sqrt{54a^2b^3} = \sqrt{3^2\cdot 6a^2b^3} = 3ab\sqrt{6b}$ $\qquad 8. \dfrac{2}{\sqrt{3}} = \dfrac{2}{\sqrt{3}}\left(\dfrac{\sqrt{3}}{\sqrt{3}}\right) = \dfrac{2\sqrt{3}}{3}$

$9. \sqrt{\dfrac{3}{17}} = \sqrt{\dfrac{3}{17}}\left(\sqrt{\dfrac{17}{17}}\right) = \dfrac{\sqrt{51}}{17}$ $\qquad 10. \sqrt[3]{\dfrac{3}{5}} = \sqrt[3]{\dfrac{3}{5}\left(\dfrac{5}{5}\right)^2} = \dfrac{\sqrt[3]{75}}{5}$

$11. \dfrac{2}{\sqrt[3]{5}} = \left(\dfrac{2}{\sqrt[3]{5}}\right)\left(\dfrac{\sqrt[3]{5^2}}{\sqrt[3]{5^2}}\right) = 2\dfrac{\sqrt[3]{25}}{5}$ $\qquad 12. \dfrac{3x\sqrt{y}}{\sqrt{2x}} = \dfrac{3x\sqrt{y}\sqrt{2x}}{\sqrt{(2x)(2x)}} = \dfrac{3x\sqrt{2xy}}{2x}$

$13. \sqrt{\dfrac{9x^3}{5x}} = \sqrt{\dfrac{9x^2}{5}} = \dfrac{3x}{\sqrt{5}}\cdot\dfrac{\sqrt{5}}{\sqrt{5}} = \dfrac{3x\sqrt{5}}{5}$ $\qquad 14. \dfrac{a}{\sqrt{a^3}} = \dfrac{a}{a\sqrt{a}} = \dfrac{1}{\sqrt{a}}\left(\dfrac{\sqrt{a}}{\sqrt{a}}\right) = \dfrac{\sqrt{a}}{a}$

$15. \sqrt{14ab^3} = b\sqrt{14ab}$ $\qquad 16. \sqrt{24xy^5} = \sqrt{(2y)^2\cdot 6xy} = 2y^2\sqrt{6xy}$

132

17. $\sqrt{\dfrac{15b}{a}} = \sqrt{\dfrac{15b}{a}\left(\dfrac{a}{a}\right)} = \dfrac{\sqrt{15ab}}{a}$ 18. $\sqrt{200xy^5} = 10y^2\sqrt{2xy}$

19. $\sqrt{8x^3y^5z^2} = 2xy^2z\sqrt{2xy}$ 20. $\dfrac{10}{\sqrt{2t}} = \dfrac{10}{\sqrt{2t}}\left(\dfrac{\sqrt{2t}}{\sqrt{2t}}\right) = \dfrac{10\sqrt{2t}}{2t} = 5\dfrac{\sqrt{2t}}{t}$

♦♦♦♦♦

6.4 OPERATIONS WITH RADICALS

OBJECTIVES:

*(A) To add and subtract radicals.
*(B) To multiply and divide radicals.

** REVIEW

*(A) To add and subtract two or more radicals we must first reduce each radical to its simplest form and combine **like terms**.
(like terms are terms which have the same index and radicand)

Examples:

a. $\sqrt{8} - \sqrt{\dfrac{1}{2}} + \sqrt{98}$

$= 2\sqrt{2} - \dfrac{\sqrt{2}}{2} + 7\sqrt{2}$ *radicand is 2 and index is 2*

$= \left(2 - \dfrac{1}{2} + 7\right)\sqrt{2}$ *factor out $\sqrt{2}$ and simplify*

$= \dfrac{17}{2}\sqrt{2}$

133

b. $\sqrt[3]{5} + \sqrt[3]{40} + 9\sqrt[5]{3} - 2\sqrt[5]{96}$

$= (\sqrt[3]{5} + 2\sqrt[3]{5}) + (9\sqrt[5]{3} - 4\sqrt[5]{3})$

$= [(1+2)\sqrt[3]{5})] + [\sqrt[5]{3}(9-4)]$

$= 3\sqrt[3]{5} + 5\sqrt[5]{3}$

c. $5\sqrt{\dfrac{4}{6}} - 3\sqrt{\dfrac{3}{8}} + 2\sqrt{\dfrac{1}{12}}$ *simplify*

$= 5\sqrt{\dfrac{2}{3}} - 3\sqrt{\dfrac{3}{2}\left(\dfrac{1}{4}\right)} + 2\sqrt{\dfrac{1}{3}\left(\dfrac{1}{4}\right)}$ *factor out* $\sqrt{\dfrac{1}{4}} = \dfrac{1}{2}$

$= 5\sqrt{\dfrac{2}{3}\left(\dfrac{3}{3}\right)} - \dfrac{3}{2}\sqrt{\dfrac{3}{2}\left(\dfrac{2}{2}\right)} + \dfrac{2}{2}\sqrt{\dfrac{1}{3}\left(\dfrac{3}{3}\right)}$ *rationalize the denominator*

$= \dfrac{5\sqrt{6}}{3} - \dfrac{3\sqrt{6}}{4} + \dfrac{\sqrt{3}}{3}$

$= \dfrac{\sqrt{6}}{12}[20-9] + \dfrac{\sqrt{3}}{3}$

$= \dfrac{11}{12}\sqrt{6} + \dfrac{\sqrt{3}}{3}$

*(B) To multiply radicals, use the <u>product rule</u> for radicals.

$$\sqrt[n]{a} \cdot \sqrt[n]{b} = \sqrt[n]{ab}$$

Examples:

a. $\sqrt{6ab} \cdot \sqrt{2a} = \sqrt{6ab \cdot 2a} = \sqrt{12a^2b} = \sqrt{(2a)^2 3b} = 2a\sqrt{3b}$

b. $\sqrt[3]{6x^2y} \cdot \sqrt[3]{4xy^5} = \sqrt[3]{6x^2y \cdot 4xy^5} = \sqrt[3]{24x^3y^6} = \sqrt[3]{(2xy^2)^3 3} = 2xy^2\sqrt[3]{3}$

134

c. $(\sqrt{5}+2)(\sqrt{5}-2) = (\sqrt{5})^2 - (2)^2 = 5-4 = 1$

d. $(3+\sqrt{x+y})(2-\sqrt{x}) = 3 \cdot 2 - 3\sqrt{x} + 2\sqrt{x+y} - \sqrt{x(x+y)}$

$$= 6 - 3\sqrt{x} + 2\sqrt{x+y} - \sqrt{x(x+y)}$$

- To divide radicals, use the **quotient rule** for radicals.

$$\frac{\sqrt[n]{a}}{\sqrt[n]{b}} = \sqrt[n]{\frac{a}{b}}$$

Examples:

a. $\dfrac{\sqrt[5]{3}}{\sqrt[5]{5}} = \sqrt[5]{\dfrac{3}{5}} = \sqrt[5]{\dfrac{3 \cdot 5^4}{5 \cdot 5^4}} = \dfrac{\sqrt[5]{3 \cdot 5^4}}{5} = \dfrac{\sqrt[5]{1875}}{5}$

b. *Simplify the radical expression.*

$$\frac{5}{2+\sqrt{3}} = \frac{5}{(2+\sqrt{3})} \cdot \frac{(2-\sqrt{3})}{(2-\sqrt{3})} = \frac{5(2-\sqrt{3})}{4-3} = 5(2-\sqrt{3})$$

$2-\sqrt{3}$ *is the conjugate of* $2+\sqrt{3}$

◆◆◆◆◆

**EXERCISES

I. Simplify the following expressions by combining like terms.

1. $\sqrt[3]{2} + \sqrt[3]{5} - 7\sqrt[3]{2}$

2. $\sqrt{xy} + 3\sqrt{xy}$

3. $\sqrt{5} + 6\sqrt{50} - 10\sqrt{5}$

4. $\dfrac{1}{\sqrt{3}} + 3\sqrt{12}$

5. $\dfrac{3}{\sqrt{5}} + \dfrac{2}{\sqrt{10}}$

6. $3\sqrt{5} - 5\sqrt{3} - \sqrt{75}$

7. $\dfrac{x}{y}\sqrt{108x^3y^2}$

8. $ab\sqrt{\dfrac{3}{2ab^2}}$

II. Perform the following operations.

9. $2\sqrt{3}\,(\sqrt{7}-2\sqrt{5})$

10. $(2\sqrt{3}+3\sqrt{2})^2$

11. $(2\sqrt{15}-5\sqrt{2})\sqrt{5}$

12. $(\sqrt{6}-2\sqrt{5})^2$

13. $(\sqrt{xy}+\sqrt{x})\sqrt{y}$

14. $(1+\sqrt{2}+\sqrt{3})\sqrt{2}$

15. $(2\sqrt{3}+\sqrt{5})(2\sqrt{3}-\sqrt{5})$

16. $5\sqrt{12}-\sqrt{108}+\sqrt{75}$

17. $\dfrac{3\sqrt{6}}{5\sqrt{3}}$

18. $\dfrac{1+\sqrt{3}}{1-\sqrt{3}}$

19. $\dfrac{3a}{b}\dfrac{\sqrt{ab^2}}{\sqrt{a^2b}}$

20. $\dfrac{-2}{2-\sqrt{3}}$

♦♦♦♦♦

**SOLUTIONS TO EXERCISES

I. Simplify the following expressions by combining like terms.

1. $\sqrt[3]{2}+\sqrt[3]{5}-7\sqrt[3]{2} = \sqrt[3]{2}-7\sqrt[3]{2}+\sqrt[3]{5} = \sqrt[3]{5}-6\sqrt[3]{2}$

2. $\sqrt{xy}+3\sqrt{xy} = 4\sqrt{xy}$

3. $\sqrt{5}+6\sqrt{50}-10\sqrt{5} = \sqrt{5}+30\sqrt{2}-10\sqrt{5} = 30\sqrt{2}-9\sqrt{5}$

4. $\dfrac{1}{\sqrt{3}}+3\sqrt{12} = \dfrac{1}{\sqrt{3}}+6\sqrt{3} = \dfrac{\sqrt{3}}{3}+6\sqrt{3} = \sqrt{3}\,(\dfrac{1}{3}+6) = \dfrac{19}{3}\sqrt{3}$

5. $\dfrac{3}{\sqrt{5}}+\dfrac{2}{\sqrt{10}} = \dfrac{3\sqrt{5}}{5}+\dfrac{2\sqrt{10}}{10} = \dfrac{3\sqrt{5}}{5}+\dfrac{\sqrt{10}}{5} = \dfrac{3\sqrt{5}+\sqrt{10}}{5}$

6. $3\sqrt{5}-5\sqrt{3}-\sqrt{75} = 3\sqrt{5}-5\sqrt{3}-5\sqrt{3} = 3\sqrt{5}-10\sqrt{3}$

7. $\dfrac{x}{y}\sqrt{108x^3y^2} = \dfrac{x}{y}\sqrt{(6xy)^2\cdot(3x)} = \dfrac{x}{y}\cdot 6xy\sqrt{(3x)} = 6x^2\sqrt{3x}$

8. $ab\sqrt{\dfrac{3}{2ab^2}} = ab(\dfrac{1}{b})\sqrt{\dfrac{3}{2a}} = a\sqrt{\dfrac{3}{2a}(\dfrac{2a}{2a})} = \dfrac{a\sqrt{6a}}{2a} = \dfrac{\sqrt{6a}}{2}$

II. Perform the following operations.

9. $2\sqrt{3}(\sqrt{7}-2\sqrt{5}) = 2\sqrt{21}-4\sqrt{15}$

10. $(2\sqrt{3}+3\sqrt{2})^2 = (2\sqrt{3})^2+2(2\sqrt{3})(3\sqrt{2})+(3\sqrt{2})^2 = 12+12\sqrt{6}+18 = 30+12\sqrt{6}$

11. $(2\sqrt{15}-5\sqrt{2})\sqrt{5} = 2\sqrt{75}-5\sqrt{10} = 10\sqrt{3}+5\sqrt{10}$

12. $(\sqrt{6}-2\sqrt{5})^2 = (\sqrt{6})^2-2(\sqrt{6})(2\sqrt{5})+(2\sqrt{5})^2 = 6-4\sqrt{30}+20 = 26-4\sqrt{30}$

13. $(\sqrt{xy}+\sqrt{x})\sqrt{y} = y\sqrt{x}+\sqrt{xy}$

14. $(1+\sqrt{2}+\sqrt{3})\sqrt{2} = \sqrt{2}+\sqrt{4}+\sqrt{6} = 2+\sqrt{2}+\sqrt{6}$

15. $(2\sqrt{3}+\sqrt{5})(2\sqrt{3}-\sqrt{5}) = (2\sqrt{3})^2-(\sqrt{5})^2 = 12-5 = 7$

16. $5\sqrt{12}-\sqrt{108}+\sqrt{75} = 10\sqrt{3}-6\sqrt{3}+5\sqrt{3} = 9\sqrt{3}$

17. $\dfrac{3\sqrt{6}}{5\sqrt{3}} = \dfrac{3\sqrt{6}}{5\sqrt{3}} \cdot \left(\dfrac{\sqrt{3}}{\sqrt{3}}\right) = \dfrac{3\sqrt{18}}{15} = \dfrac{\sqrt{9\cdot 2}}{5} = \dfrac{3\sqrt{2}}{5}$

18. $\dfrac{1+\sqrt{3}}{1-\sqrt{3}} = \left(\dfrac{1+\sqrt{3}}{1-\sqrt{3}}\right) \cdot \left(\dfrac{1+\sqrt{3}}{1+\sqrt{3}}\right) = \dfrac{1+2\sqrt{3}+3}{1-3} = \dfrac{4+2\sqrt{3}}{-2} = -(2+\sqrt{3})$

19. $\dfrac{3a}{b}\dfrac{\sqrt{ab^2}}{\sqrt{a^2b}} = \dfrac{3ab\sqrt{a}}{ab\sqrt{b}} = 3\sqrt{\dfrac{a}{b}} = 3\dfrac{\sqrt{ab}}{b}$

20. $\dfrac{-2}{2-\sqrt{3}} = \left(\dfrac{-2}{2-\sqrt{3}}\right) \cdot \left(\dfrac{2+\sqrt{3}}{2+\sqrt{3}}\right) = -\dfrac{(4+2\sqrt{3})}{4-3} = -(4+2\sqrt{3})$

◆◆◆◆◆

6.5 SOLVING EQUATIONS WITH RADICALS

OBJECTIVES:

*(A) To solve $x^2 = k$ using the square root property.
*(B) To solve for the radical variable by the quotient rule.

** REVIEW

*(A) Consider the equation $x^2 = k$
     If $k > 0$, $x = \pm \sqrt{k}$.
     If $k = 0$, $x = 0$.
     If $k < 0$, x has no real root.

Examples:

     a. $x^2 = 10$ b. $3(x-1)^2 = 27$ c. $(x-5)^2 = 0$

Solutions:

     a. $x^2 = 10$ b. $3(x-1)^2 = 27$ c. $(x-5)^2 = 0$
        $x = \pm \sqrt{10}$ $(x-1)^2 = 9$ $x-5 = 0$
                           $x - 1 = \pm 3$ $x = 5$
                           $x = 4$ or -2

- In order to solve an equation, it is sometimes necessary to square each side of an equation. However the new equation may have extraneous roots.

Example:

$$\sqrt{2 - x} = x$$
$$2 - x = x^2 \qquad \text{square both sides of the equation}$$
$$x^2 + x - 2 = 0 \qquad \text{rearrange the terms}$$
$$(x + 2)(x - 1) = 0 \qquad \text{factor the equation}$$
$$x = -2 \text{ or } x = 1$$

check:
  If $x = -2$ If $x = 1$

$\sqrt{2 - x}$ $\sqrt{2 - x}$
$= \sqrt{2-(-2)}$ $= \sqrt{2 - (1)}$
$= \sqrt{4}$ $= \sqrt{1}$
$= 2$ which is not equal to x. $= 1$ which is equal to x.

$x = 1$ is the solution of the equation and $x = -2$ is an extraneous root.

*(B) To solve for an indicated variable:

     Step 1. If the equation involves radicals, isolate the radical on one side of the equation.

     Step 2. Rationalize both sides of the equation.

     Step 3. Solve the equation and check for extraneous root(s).

Examples:

    a. $a^2 - b = 0$ solve for a

    b. $\sqrt{3xy} - 2 = 0$ solve for y

Solutions:

    a. $a^2 - b = 0$ -- isolate the variable a on one side of the equation

       $a^2 = b$

       $a = \pm \sqrt{b}$

    b. $\sqrt{3xy} - 2 = 0$

       $\sqrt{3xy} = 2$ -- isolate the radical on one side of the equation

       $3xy = 4$ -- square both sides of the equation and solve for y

       $y = 4/(3x)$

check:

When $a = \pm\sqrt{b}$ When $y = 4/(3x)$

  $a^2 - b$ $\sqrt{3xy} - 2$

  $= (\pm\sqrt{b})^2 - b$ $= \sqrt{4} - 2$

  $= b - b$ $= 0$ correct

  $= 0$ correct

                                 ♦♦♦♦♦

EXERCISES

I. Solve each equation.

1. $x^2 = 9$ 2. $x^2 = 121$ 3. $x^2 = 10$

4. $5x^2 = 4$ 5. $12x^2 = 6$ 6. $(x-3)^2 = 5$

7. $(x+10)^2 = 0$ 8. $\sqrt{x-8} = 1$ 9. $\sqrt{6x-5} = -1$

10. $\sqrt{3x+4} = 2$ 11. $\sqrt{x+12} = x$ 12. $\sqrt{x+5} = \sqrt{x-5}$

13. $\sqrt{3x+5} = \sqrt{2x-3}$ 14. $x = \sqrt{10-3x}$ 15. $\sqrt{x} = \sqrt{5-4x}$

II. *Solve for the indicated variable.*

16. $x^2+y^2=1$ *for x*   17. $a^2-b^2-c=0$ *for b* 18. $gt^2=l\pi^2$ *for t*

19. $m=\sqrt{3ptr}$ *for r*   20. $x^3=y+1$ *for x*

♦♦♦♦♦

SOLUTIONS TO EXERCISES

I. Solve each equation.

1. $x^2 = 9$ 2. $x^2 = 121$ 3. $x^2 = 10$

   $x = \pm 3$ $x = \pm 11$ $x = \pm\sqrt{10}$

4. $5x^2 = 4$ 5. $12x^2 = 6$ 6. $(x - 3)^2 = 5$

   $x = \pm 2/\sqrt{5}$ $x = \pm 1/\sqrt{2}$ $x - 3 = \pm\sqrt{5}$

   $x = \pm (2\sqrt{5})/5$ $x = \pm\sqrt{2}/2$ $x = 3 \pm \sqrt{5}$

7. $(x + 10)^2 = 0$ 8. 8. $\sqrt{x-8} = 1$ 9. $\sqrt{6x-5} = -1$

   $x + 10 = 0$ $x - 8 = 1$ There is no real root.

   $x = -10$ $x = 9$

140

10. $\sqrt{3x+4} = 2$
$3x+4 = 4$
$3x = 0$
$x=0$

11. $\sqrt{x+12} = x$
$x+12 = x^2$
$x^2-x-12 = 0$
$(x-4)(x+3)=0$
$x=4$ ($x=-3$ *is not a root*).

12. $\sqrt{x+5} = \sqrt{x-5}$
$x+5 = x-5$
no solution

13. $\sqrt{3x+5} = \sqrt{2x-3}$
$3x+5 = 2x-3$
($x=-8$ *is not a root*).

14. $x = \sqrt{10-3x}$
$x^2 = 10-3x$
$x^2+3x-10 = 0$
$(x-2)(x+5) = 0$
$x=2$ ($x=-5$ *is not a root*).

15. $\sqrt{x} = \sqrt{5-4x}$
$x = 5-4x$
$5x = 5$
$x = 1$

II.
Solve for the indicated variable.

16. $x^2+y^2 = 1$ *for* x
$x^2 = 1-y^2$
$x = \pm\sqrt{1-y^2}$

17. $a^2-b^2-c = 0$ *for* b
$b^2 = a^2-c$
$b = \pm\sqrt{a^2-c}$

18. $gt^2 = l\pi^2$ *for* t
$t^2 = \dfrac{l}{g}\pi^2$

$t = \pm\sqrt{\dfrac{l}{g}}\,\pi$

19. $m = \sqrt{3ptr}$ *for* r
$m^2 = 3ptr$

$r = \dfrac{m^2}{3pt}$

20. $x^3 = y+1$ *for* x
$x = \sqrt[3]{y+1}$

◆◆◆◆◆

6.6 NEGATIVE AND RATIONAL EXPONENTS

OBJECTIVES:

*(A) To understand negative integral exponents.
*(B) To understand rational exponents.

** REVIEW

*(A) Negative integral exponents

If a is a nonnegative real number and n is a positive integer, then

$$a^{-n} = \frac{1}{a^n} \quad and \quad a^n = \frac{1}{a^{-n}}$$

For any integer (positive or negative) m and n and a ≠ 0, b ≠ 0, then

- **Product Rule:**

  $a^m \cdot a^n = a^{(m+n)}$

- **Quotient Rule:**

  $\dfrac{a^m}{a^n} = a^{m-n}$

- **The Power Rule:**

  $(a^m)^n = a^{mn}$ *power rule*

  $(ab)^n = a^n b^n$ *power of a product*

  $\left(\dfrac{a}{b}\right)^n = \dfrac{a^n}{b^n}$ *power of a quotient*

  $b^{\frac{1}{n}} = \sqrt[n]{b}$ *n^{th} root of b*

Examples:

Simplify the following

a.

$(x^2)^{-3}$

$= x^{2 \cdot (-3)}$ *power rule*

$= x^{-6}$

$= \dfrac{1}{x^6}$

b.

$\left(\dfrac{3a^{-2}}{a^4}\right)^{-3}$

$= \dfrac{3^{-3}a^{(-2)(-3)}}{(a^4)^{-3}}$

$= \dfrac{a^6}{3^3 a^{-12}} = \dfrac{1}{27}a^{6-(-12)}$

$= \dfrac{1}{27}a^{18}$

*(B) Rational exponents.

If m and n are positive integers, then

(a) For positive fractional exponents:

$a^{(m/n)} = (a^{1/n})^m$ provided $a^{1/n}$ is a real number.

(b) For negative fractional exponents:

$a^{\left(-\frac{m}{n}\right)} = \dfrac{1}{a^{\left(\frac{m}{n}\right)}}$,   *provide* $a^{\frac{1}{n}}$ *is a real and* $a \neq 0$

Consider: $a^{\left(\frac{m}{n}\right)} = \sqrt[n]{a^m}$ *The expression means* ---

Raise a to the mth power and find n^{th} root of $(a)^m$

Examples:

$$a. \quad a^{-\frac{2}{5}} = \frac{1}{a^{\frac{2}{5}}} = \frac{1}{\sqrt[5]{a^2}}$$

$$b. \quad 81^{\frac{-3}{4}} = ((81)^{\frac{1}{4}})^{-3}$$

$$= ((3^4)^{\frac{1}{4}})^{-3} = 3^{-3} = \frac{1}{27}$$

$$c. \quad (-8)^{-\frac{2}{3}} = ((-2)^3)^{\frac{-2}{3}} = (-2)^{-2} = \frac{1}{4}$$

♦♦♦♦♦

*EXERCISES

I. Evaluate each expression.

1. $(-2)^3$

2. 2^{-3}

3. $-(5)^{-2}$

4. $(\frac{2^{-3}}{16^{-2}})$

5. $\frac{10^{-2}}{10^{-3}}$

6. $(\frac{3}{2})^{-1}$

II. Simplify each of the following:

Write the result with   positive exponents only.

7. $(a^{-3})^5$

8. $(x^{-2})^{-1}$

9. $a^{-2} \cdot a^2$

10. $(\frac{a^{-2}}{a^{-6}})^{-1}$

11. $(\frac{2x^3}{3x^5})^{-2}$

12. $-(a)^{-\frac{2}{3}}$

13. $x^{\frac{1}{3}} x^{\frac{1}{3}} x^{\frac{1}{3}}$

14. $(-1)^{\frac{3}{5}}$

15. $(-x^3)^{-\frac{2}{3}}$

16. $\frac{3x^0}{3^{-2}}$

17. $(a^{-2})^{-1} \cdot (-a)^{-1}$

18. $a^{4p} \cdot a^{-4p}$

19. $(\frac{a^{-2}}{b^{-5}})^{\frac{5}{2}}$

20. $((\frac{a}{b})^{-\frac{5}{2}})^{-\frac{2}{5}}$

♦♦♦♦♦

**SOLUTIONS TO EXERCISES

1. $(-2)^3 = -8$

2. $2^{-3} = \dfrac{1}{2^3} = \dfrac{1}{8}$

3. $-(5)^{-2} = -\dfrac{1}{25}$

4. $\left(\dfrac{2^{-3}}{16^{-2}}\right) = \dfrac{16^2}{2^3} = \dfrac{(2^4)^2}{2^3} = \dfrac{2^8}{2^3}$

$= 2^{8-3} = 32$

5. $\dfrac{10^{-2}}{10^{-3}} = \dfrac{10^3}{10^2} = 10$

6. $\left(\dfrac{3}{2}\right)^{-1} = \dfrac{2}{3}$

II. Simplify each expression.

Write your answer with positive exponents only.

7. $(a^{-3})^5 = a^{(-3)\cdot 5} = a^{-15} = \dfrac{1}{a^{15}}$

8. $(x^{-2})^{-1} = x^{(-2)(-1)} = x^2$

9. $a^{-2} \cdot a^2 = a^{-2+2} = a^0 = 1$

10. $\left(\dfrac{a^{-2}}{a^{-6}}\right)^{-1} = (a^{-2-(-6)})^{-1}$

$= (a^4)^{-1} = a^{-4} = \dfrac{1}{a^4}$

11. $\left(\dfrac{2x^3}{3x^5}\right)^{-2} = \left(\dfrac{3x^5}{2x^3}\right)^2 = \dfrac{(3x^5)^2}{(2x^3)^2}$

$= \dfrac{3^2 \cdot x^{10}}{2^2 \cdot x^6} = \dfrac{9}{4}x^4$

12. $-(a)^{-\frac{2}{3}} = -\dfrac{1}{a^{\frac{2}{3}}} = -\dfrac{a^{\frac{1}{3}}}{a}$

13. $x^{\frac{1}{3}}x^{\frac{1}{3}}x^{\frac{1}{3}} = x^{\frac{1}{3}+\frac{1}{3}+\frac{1}{3}} = x$

14. $(-1)^{\frac{3}{5}} = \sqrt[5]{(-1)^3} = -1$

15. $(-x^3)^{-\frac{2}{3}} = \left((-x^3)^{\frac{1}{3}}\right)^{-2} = (-x)^{-2} = \frac{-1}{x^2}$

16. $\frac{3x^0}{3^{-2}} = 3\cdot3^2 = 27$

17. $(a^{-2})^{-1}\cdot(-a)^{-1} = a^{(-2)(-1)}\cdot(-a)^{-1}$

$\quad = a^2(-a)^{-1} = a^2\left(\frac{-1}{a}\right) = -a$

18. $a^{4p}\cdot a^{-4p} = a^{4p-4p} = a^0 = 1$

19. $\left(\frac{a^{-2}}{b^{-5}}\right)^{\frac{1}{5}} = \left(\frac{b^5}{a^2}\right)^{\frac{1}{5}} = \frac{b^{5\cdot\frac{1}{5}}}{a^{2\cdot\frac{1}{5}}} = \frac{b}{a^{\frac{2}{5}}} = \frac{a^{\frac{3}{5}}b}{a}$

20. $\left(\left(\frac{a}{b}\right)^{-\frac{5}{2}}\right)^{-\frac{2}{5}} = \left(\frac{a}{b}\right)^{(-\frac{5}{2})\cdot(\frac{-2}{5})} = \frac{a}{b}$

♦♦♦♦♦

6.7 SCIENTIFIC NOTATION

OBJECTIVES:

*(A) To convert scientific notation to standard notation and vice versa.
*(B) To perform computations using scientific notation.

** REVIEW

*(A) Scientific notation is a number written as a product of a number between 1 and 10, and a power of 10.
There is **one nonzero digit** to the left of the decimal point.

Examples:

These numbers are all in scientific notation.

a. 3.5×10^5 b. 2.0×10^{-3} c. 1.9×10^6

• To convert from scientific notation to standard notation.

(1). If the power of 10 is a positive integer n, simply move the decimal point n places to the right.

Example:

$\qquad 3.5 \times 10^4 \qquad$ n equals 4, therefore move
$\qquad\qquad\qquad\qquad$ the decimal point 4 places to the right
$\qquad 3.5 \times 10^4 = 35000$

(2). If the power of 10 is a negative integer n, simply move the decimal point n places to the left.

Examples:

    a. $8.1 \times 10^{-2} = 0.081$ (n equals -2, therefore move the decimal point 2 places to the left)

    b. $7.0 \times 10^{-5} = 0.00007$ (n equals -5, therefore move the decimal point 5 places to the left)

- To convert a number from standard notation to scientific notation, reverse the above procedure.

Examples:

    a. 723000.   (move the decimal point 5 places to the left and multiply the number by 10^5)

    $723000 = 7.23 \times 10^5$

    b. 0.00235   (move the decimal point 3 places to the right and multiply the number by 10^{-3})

    $0.00235 = 2.35 \times 10^{-3}$

◆◆◆◆◆

**EXERCISES

**I. Write each of the following numbers in standard notation.

    1. 3×10^3 2. 3.05×10^1 3. 1×10^7
    4. 0.00012×10^6 5. 0.2×10^4 6. 3101×10^2

**II. Write each of the following numbers in scientific notation.

    7. 500,000   8. 21000000000   9. 750
  10. 60125   11. 2   12. 420×10^5
  13. 920×10^{-5} 14. $(0.002)(0.3)$ 15. 0.00023
  16. 0.00023×10^{-1} 17. $(0.0002)^3$ 18. 0.325×10^{-5}
  19. 0.00005   20. 123456.123

**SOLUTIONS TO EXERCISES

**I. Write each of the following numbers in standard notation.

    1. $3 \times 10^3 = 3 \times 1000 = 3,000$

    2. $3.05 \times 10^1 = 3.05 \times 10 = 30.5$

    3. $1 \times 10^7 = 10,000,000$

    4. $0.00012 \times 10^6 = 0.00012 \times 1000000 = 120$

    5. $0.2 \times 10^4 = 0.2 \times 10000 = 2000$

    6. $3101 \times 10^2 = 3101 \times 100 = 310100$

**II. Write each of the following numbers in scientific notation.

    7. $500,000 = 500,000.0 = 5 \times 10^5 = 5.0 \times 10^5$

      (move the decimal point 5 places to the left

       and multiply the number by 10^5)

    8. $21000000000 = 2,100,000,000.0 = 21 \times 10^8 = 2.1 \times 10^9$

      (move the decimal point 9 places to the left

       and multiply the number by 10^9)

    9. $750 = 750.0 = 7.5 \times 10^2$

  10. $60125 = 60,125.0 = 6.0125 \times 10^4$

  11. $2 = 2.0 \times 10^0$

12. $420 \times 10^5 = 4.2 \times 10^2 \times 10^5 = 4.2 \times 10^7$

13. $920 \times 10^{-5} = 9.2 \times 10^2 \times 10^{-5} = 9.2 \times 10^{2-5} = 9.2 \times 10^{-3}$

14. $(0.002)(0.3) = 0.0006 = 6.0 \times 10^{-4}$

15. $0.00023 = 2.3 \times 10^{-4}$

16. $0.00023 \times 10^{-1} = 2.3 \times 10^{-4} \times 10^{-1} = 2.3 \times 10^{-4-1} = 2.3 \times 10^{-5}$

17. $(0.0002)^3 = (2.0 \times 10^{-4})^3 = (2.0)^3 (10^{-4})^3 = 8.0 \times 10^{-12}$

18. $0.325 \times 10^{-5} = (3.25 \times 10^{-1}) \times 10^{-5} = 3.25 \times 10^{-1-5} = 3.25 \times 10^{-6}$

19. $0.00005 = 5.0 \times 10^{-5}$
    (move the decimal point 5 places to the right
    and multiply by 10^{-5})

20. $123456.123 = 1.23456123 \times 10^5$
    (move the decimal point 5 places to the left
    and multiply by 10^5)

♦♦♦♦♦

CHAPTER VI - TEST A

Simplify the following.

1. $(-3)^4$

2. -3^4

3. $a^{-5}a^8$

4. $5^0\,5^{-2}$

5. $(a^{-2})^3$

6. $(x^5)^{-2}$

7. 10^{-2}

8. $(40)^2$

9. $\sqrt[3]{(-8a^6)}$

10. $\sqrt{10000}$

11. $\sqrt[3]{(-27a^9)}$

12. $\sqrt{25x^2}$

13. $\sqrt{12a^3x}$

14. $(-\tfrac{1}{2}x^2y)^3$

15. $(3^0 - 5)^0$

16. $(3)^{-5})(3)^3$

17. $(10)^{-2}$

18. $3x^{-2}\cdot 4x^5$

19. $2a^3b(a^3)^{-2}(b^4)^{-3}$

20. $(ab^2)^3(b^2a)^{-3}$

♦♦♦♦♦

150

I. Simplify the following expressions.
   (Use the product rule and/or quotient rule for radicals)

1. $\sqrt{k^4}$

2. $\sqrt[3]{a^9}$

3. $\sqrt[7]{3^{49}}$

4. $\sqrt[4]{x^8}$

5. $\sqrt[5]{\dfrac{32a^{10}}{b^{15}}}$

6. $\sqrt[3]{\dfrac{a}{9}}$

7. $\sqrt[3]{\dfrac{a^4}{27}}$

8. $\sqrt[5]{100000a^{10}}$

9. $\sqrt[3]{\dfrac{a^2b^4}{64}}$

II. *Solve each equation.*

10. $x^3 = -8$

11. $3x^2 = 4$

12. $8x^2 = 16$

13. $(x-1)^2 = 2$

14. $x = \sqrt{10-3x}$

15. $x-1 = \sqrt{x+1}$

III. Perform the computations and simplify.

16. $(2\sqrt{3}-4\sqrt{2})^2$

17. $(\sqrt{7}-\sqrt{8})(\sqrt{7}+\sqrt{8})$

18. $\dfrac{\sqrt{5}}{\sqrt{3}-\sqrt{4}}$

19. $\dfrac{3-\sqrt{5}}{2\sqrt{5}}$

20. $\dfrac{2}{\sqrt{3}-1}$

♦♦♦♦♦

CHAPTER VI - TEST C

I. Solve each equation.

1. $\sqrt{3x-5} = \sqrt{5x-9}$

2. $x+1 = \sqrt{2x-1}$

3. $5(x-2)^2 - 6(x-2) + 1 = 0$

4. $\sqrt{9x-5} = 0$

5. $\left(\frac{1}{2}x+3\right)^2 = 0$

6. $4(x-3)^3 + 32 = 0$

II. Simplify the following expression by combining like radicals.

7. $\sqrt{3} + 5\sqrt{3}$

8. $\sqrt[3]{7} - \sqrt[3]{8} + 3\sqrt[3]{6}$

9. $9\sqrt{a} - 8\sqrt{b} + 2\sqrt{b} + \sqrt{a}$

10. $\sqrt{28} + 3\sqrt{7} - \sqrt{63}$

11. $\sqrt{27a^3} - \sqrt{8a}$

12. $\sqrt{20} + \sqrt{45} + \sqrt{500}$

III. Simplify each expression.

13. $\dfrac{3}{\sqrt{2}-\sqrt{3}}$

14. $\dfrac{1}{\sqrt{3}-5}$

15. $\dfrac{\sqrt{2}+\sqrt{3}}{\sqrt{4}}$

16. $\dfrac{2-\sqrt{3}}{2+\sqrt{3}}$

17. $\dfrac{-3}{\sqrt{5}-\sqrt{4}}$

18. $\dfrac{\sqrt{2}-\sqrt{5}}{\sqrt{2}}$

19. $\dfrac{\sqrt{3}}{\sqrt{27}}$

20. $\dfrac{9-\sqrt{4}}{2}$

◆◆◆◆◆

152

CHAPTER VI - TEST A - SOLUTIONS

Simplify the following.

1. $(-3)^4 = (-3)(-3)(-3)(-3) = 81$

2. $-3^4 = -(3)(3)(3)(3) = -81$

3. $a^{-5}a^8 = a^{-5+8} = a^3$ 4. $5^0 5^{-2} = 5^{0-2} = 5^{-2} = 1/25$

5. $(a^{-2})^3 = a^{-6} = 1/(a^6)$ 6. $(x^5)^{-2} = x^{-10} = 1/x^{10}$

7. $10^{-2} = 1/100$ 8. $(40)^2 = 1600$

9. $\sqrt[3]{(-8a^6)} = \sqrt[3]{(-2a^2)^3} = -2a^2$ 10. $\sqrt{10000} = 100$

11. $\sqrt[3]{(-27a^9)} = \sqrt[3]{(-3a^3)^3} = -3a^3$ 12. $\sqrt{25x^2} = 5x$

13. $\sqrt{12a^3} = \sqrt{(4a^2)(3a)} = 2a\sqrt{3a}$ 14. $(-\tfrac{1}{2}x^2y)^3 = -(1/8)x^6y^3$

15. $(3^0 - 5)^0 = (1 - 5)^0 = (-4)^0 = 1$

16. $3^{-5} \cdot (3^3) = = 3^{-5+3} = 3^{-2} = 1/9$

17. $(10)^{-2} = 1/10^2 = 1/100$

18. $3x^{-2} \cdot 4x^5 = 12x^{-2+5} = 12x^3$

19. $2a^3b(a^3)^{-2}(b^4)^{-3} = 2a^3a^{-6}b^1b^{-12}$

    $= 2a^{3-6}b^{1-12} = 2a^{-3}b^{-11} = 2/(a^3b^{11})$

20. $(ab^2)^3(b^2a)^{-3} = a^3b^6b^{-6}a^{-3} = a^{3-3}b^{6-6} = a^0b^0 = 1$

♦♦♦♦♦

CHAPTER VI - TEST B - SOLUTIONS

1. $\sqrt{k^4}=k^2$

2. $\sqrt[3]{a^9}=\sqrt[3]{(a^3)^3}=a^3$

3. $\sqrt[7]{3^{49}}=\sqrt[7]{(3^7)^7}=3^7=2187$

4. $\sqrt[4]{x^8}=\sqrt[4]{(x^4)^2}=x^2$

5. $\sqrt[5]{\dfrac{32a^{10}}{b^{15}}}=\sqrt[5]{\dfrac{(2a^2)^5}{(b^3)^5}}=\dfrac{2a^2}{b^3}$

6. $\sqrt[3]{\dfrac{a}{9}}=\sqrt[3]{\dfrac{a}{9}}\cdot\sqrt[3]{\dfrac{3}{3}}=\dfrac{\sqrt[3]{3a}}{3}$

7. $\sqrt[3]{\dfrac{a^4}{27}}=\dfrac{a\sqrt[3]{a}}{3}$

8. $\sqrt[5]{100000a^{10}}=\sqrt[5]{(10a^2)^5}=10a^2$

9. $\sqrt[3]{\dfrac{a^2b^4}{64}}=\dfrac{b\sqrt[3]{a^2b}}{4}$

II.

Solve each equation.

10. $x^3=-8$

    $x^3=(-2)^3$

    $x=-2$

11. $3x^2=4$

    $x^2=\dfrac{4}{3}$

    $x=\pm\dfrac{2}{\sqrt{3}}$

12. $8x^2=16$

    $x^2=2$

    $x=\pm\sqrt{2}$

13. $(x-1)^2=2$

    $(x-1)=\pm\sqrt{2}$

    $x=\pm\sqrt{2}+1$

14. $x=\sqrt{10-3x}$

    $x^2=10-3x$

    $x^2+3x-10=0$

    $(x-2)(x+5)=0$

    $x=2$

    (discard x=-5)

15. $x-1=\sqrt{x+1}$

    $x^2-2x+1=x+1$

    $x^2-3x=0$

    $x(x-3)=0$

    $x=3$

    (discard x=0).

16. $(2\sqrt{3}-4\sqrt{2})^2=(2\sqrt{3})^2-2(2\sqrt{3})(4\sqrt{2})+(4\sqrt{2})^2$

$=12-16\sqrt{6}+32$

$=44-16\sqrt{2}$

$=4(11-4\sqrt{2})$

17. $(\sqrt{7}-\sqrt{8})(\sqrt{7}+\sqrt{8})=(\sqrt{7})^2-(\sqrt{8})^2=7-8=-1$

18. $\dfrac{\sqrt{5}}{\sqrt{3}-\sqrt{4}}=\dfrac{\sqrt{5}(\sqrt{3}+\sqrt{4})}{(\sqrt{3}-\sqrt{4})(\sqrt{3}+\sqrt{4})}=\dfrac{\sqrt{5}(\sqrt{4}+\sqrt{3})}{3-4}=-\sqrt{5}(\sqrt{4}+\sqrt{3})$

19. $\dfrac{3-\sqrt{5}}{2\sqrt{5}}=\dfrac{(3-\sqrt{5})\sqrt{5}}{(2\sqrt{5})\sqrt{5}}=\dfrac{\sqrt{5}(3-\sqrt{5})}{10}$

20. $\dfrac{2}{\sqrt{3}-1}=\dfrac{2}{\sqrt{3}-1}\cdot\dfrac{\sqrt{3}+1}{\sqrt{3}+1}=\dfrac{2(\sqrt{3}+1)}{3-1}=\sqrt{3}+1$

♦♦♦♦♦

CHAPTER VI - TEST C - SOLUTIONS

I. Solve each equation.

1. $\sqrt{3x-5} = \sqrt{5x-9}$

   $3x-5 = 5x-9$

   $-2x = -4$

   $x = 2$

2. $x+1 = \sqrt{2x-1}$

   $(x+1)^2 = 2x+1$

   $x^2+2x+1 = 2x+1$

   $x = 0$ (*extraneous root*)

3. $5(x-2)^2-6(x-2)+1 = 0$

   Let $a = (x-2)$

   $5a^2-6+1 = 0$

   $(5a-1)(a-1) = 0$

   $a = \frac{1}{5}, \ a = 1$

   $(x-2) = \frac{1}{5}$ *or* $(x-2) = 1$

   $x = \frac{11}{5}, \ x=3$

4. $\sqrt{9x-5} = 0$

   $9x-5 = 0$

   $9x = 5$

   $x = \frac{5}{9}$

5. $\left(\frac{1}{2}x+3\right)^2 = 0$

   $\frac{1}{2}x+3 = 0$

   $x+6 = 0$

   $x = -6$

6. $4(x-3)^3+32 = 0$

   $(x-3)^3+8 = 0$

   $(x-3)^3 = -8$

   $(x-3) = -2$

   $x = 1$

II. Simplify the following expression by combining like radicals.

7. $\sqrt{3}+5\sqrt{3}=6\sqrt{3}$

8. $\sqrt[3]{7}-\sqrt[3]{8}+3\sqrt[3]{6}=\sqrt[3]{7}-2+3\sqrt[3]{6}$

9. $9\sqrt{a}-8\sqrt{b}+2\sqrt{b}+\sqrt{a}=10\sqrt{a}-6\sqrt{b}$

10. $\sqrt{28}+3\sqrt{7}-\sqrt{63}=2\sqrt{7}+3\sqrt{7}-3\sqrt{7}=2\sqrt{7}$

11. $\sqrt{27a^3}-\sqrt{8a}=3a\sqrt{3a}-2\sqrt{2a}$

12. $\sqrt{20}+\sqrt{45}+\sqrt{500}=2\sqrt{5}+3\sqrt{5}+10\sqrt{5}=15\sqrt{5}$

◆◆◆◆◆

III. Simplify each expression.

13. $\dfrac{3}{\sqrt{2}-\sqrt{3}} = \dfrac{3\left(\sqrt{2}+\sqrt{3}\right)}{\left(\sqrt{2}-\sqrt{3}\right)\left(\sqrt{2}+\sqrt{3}\right)} = \dfrac{3\left(\sqrt{2}+\sqrt{3}\right)}{2-3} = -3\left(\sqrt{2}+\sqrt{3}\right)$

14. $\dfrac{1}{\sqrt{3}-5} = \dfrac{1}{\sqrt{3}-5}\cdot\dfrac{\sqrt{3}+5}{\sqrt{3}+5} = \dfrac{\sqrt{3}+5}{3-25} = -\dfrac{1}{22}\left(\sqrt{3}+5\right)$

15. $\dfrac{\sqrt{2}+\sqrt{3}}{\sqrt{4}} = \dfrac{\sqrt{2}+\sqrt{3}}{2}$

16. $\dfrac{2-\sqrt{3}}{2+\sqrt{3}} = \dfrac{\left(2-\sqrt{3}\right)}{\left(2+\sqrt{3}\right)}\cdot\dfrac{\left(2-\sqrt{3}\right)}{\left(2-\sqrt{3}\right)} = \dfrac{4-4\sqrt{3}+3}{4-3} = 7-4\sqrt{3}$

17. $\dfrac{-3}{\sqrt{5}-\sqrt{4}} = \dfrac{-3}{\left(\sqrt{5}-\sqrt{4}\right)}\cdot\dfrac{\sqrt{5}+\sqrt{4}}{\sqrt{5}+\sqrt{4}} = \dfrac{-3\left(\sqrt{5}+\sqrt{4}\right)}{5-4} = -3\left(\sqrt{5}+2\right)$

18. $\dfrac{\sqrt{2}-\sqrt{5}}{\sqrt{2}} = \dfrac{\left(\sqrt{2}-\sqrt{5}\right)}{\sqrt{2}}\cdot\dfrac{\sqrt{2}}{\sqrt{2}} = \dfrac{1}{2}\left(2-\sqrt{10}\right)$

19. $\dfrac{\sqrt{3}}{\sqrt{27}} = \dfrac{\sqrt{3}}{3\sqrt{3}} = \dfrac{1}{3}$

20. $\dfrac{9-\sqrt{4}}{2} = \dfrac{9-2}{2} = \dfrac{7}{2}$

♦♦♦♦♦

157

CHAPTER 7

QUADRATIC EQUATIONS

7.1   __FAMILIAR QUADRATIC EQUATIONS__

__OBJECTIVES__:

*(A) To introduce the simplest quadratic equation.
*(B) To solve the quadratic equation by factoring.

** __REVIEW__

*(A) A quadratic equation has the form $ax^2 + bx + c = 0$ where a,b and c
     are constants ($a \neq 0$). If $b = 0$ in $ax^2 + bx + c = 0$,
     then $ax^2 + c = 0$ is called a simple quadratic equation.

Examples:

     $x^2 - 16 = 0$, $-6x^2 + 5 = 0$ are simple quadratic equations.

● To solve the simple quadratic equation, isolate the x^2 term then take
  the square root of both sides to find the solutions.

    Examples:

a. $2x^2 - 8 = 0$		b. $3(x-3)^2 = 51$
$2x^2 = 8$ <--- isolate x^2 term ---->		$(x-3)^2 = 17$
$x^2 = 4$ <--- take the square root -->		$x - 3 = \pm\sqrt{17}$
$x = \pm 2$ of both sides		$x = 3 \pm \sqrt{17}$

*(B) To solve quadratic equations by factoring, do the following:

    1. Isolate all the terms on the left side of the equality
       so that the right side is zero.
    2. Factor the left side of the equation.
    3. Set each factor equal to zero and find the x values.
    4. Check the answers in the original equation.

Examples:

     $2x^2 - 9x = 5$ --- isolate all the terms on the left so that
     $2x^2 - 9x - 5 = 0$ --- right side of the equation is 0.
     $(2x+1)(x-5) = 0$ --- factor the quadratic equation
     $2x+1 = 0$ or $x - 5 = 0$ --- apply the zero factor property
     $x = -\frac{1}{2}$ or $x = 5$ --- solve for x

Check:
     If $x = -\frac{1}{2}$, $2x^2 - 9x = 2(-\frac{1}{2})^2 - 9(-\frac{1}{2}) = \frac{1}{2} + 9/2 = 5$.
     If $x = 5$, $2x^2 - 9x = 2(5)^2 - 9(5) = 5$.
     Both $x = -\frac{1}{2}$ and $x = 5$ check, so the solution is $-\frac{1}{2}$ or 5.

                     ◆◆◆◆◆

I. Solve the following quadratic equations.

1. $x^2 - 12 = 0$ 2. $x^2 - 1 = 0$ 3. $x^2 + 1 = 0$

4. $3x^2 - 15 = 0$ 5. $(x-2)^2 = 9$ 6. $(2n-5)^2 = 9$

7. $(2x-1)^2 = 1/9$ 8. $3x^2 + 6x + 3 = 0$ 9. $-9x^2 + 5x = 0$

10. $6x^2 + 19x + 15 = 0$ 11. $81x^2 - 18x + 1 = 0$

12. $x^2 - 11x + 28 = 0$ 13. $x^2 = 4x$ 14. $x^2 = 9$

15. $x^2 - x - 12 = 0$ 16. $x^2 - 14x + 49 = 0$ 17. $\frac{1}{2}(x-2)^2 = 2$

18. $6x - 3 = 5 + x^2$ 19. $(3x - 5)^2 - 5 = 0$ 20. $\sqrt{2x+5} = \sqrt{x^2+2}$

◆◆◆◆◆

SOLUTIONS TO EXERCISES

I. Solve the following quadratic equations.

1. $x^2 - 12 = 0$
   $x^2 = 12$
   $x = \pm 2\sqrt{3}$

2. $x^2 - 1 = 0$
   $x^2 = 1$
   $x = \pm 1$

3. $x^2 + 1 = 0$
   $x^2 = -1$
   $x = \pm i$

4. $3x^2 - 15 = 0$
   $x^2 = 5$
   $x = \pm \sqrt{5}$

5. $(x-2)^2 = 9$
   $(x-2) = \pm 3$
   $x = 5, -1$

6. $(2n-5)^2 = 9$
   $(2n-5) = \pm 3$
   $n = 1, 4$

7. $(2x - 1)^2 = 1/9$
   $2x - 1 = \pm(1/3)$
   $x = 2/3, 1/3$

8. $3x^2 + 6x + 3 = 0$
   $3(x^2 + 2x + 1) = 0$
   $3(x+1)^2 = 0$
   $x = -1$

9. $-9x^2 + 5x = 0$
   $x(-9x+5) = 0$
   $x = 0, 5/9$

10. $6x^2 + 19x + 15 = 0$
    $(2x+3)(3x+5) = 0$
    $x = -3/2, -5/3$

11. $81x^2 - 18x + 1 = 0$
    $(9x-1)^2 = 0$
    $x = 1/9$

12. $x^2 - 11x + 28 = 0$
    $(x-7)(x-4) = 0$
    $x = 4, 7$

13. $x^2 = 4x$
    $x = 0, 4$

14. $x^2 = 9$
    $x = \pm 3$

15. $x^2 - x - 12 = 0$
    $(x+3)(x-4) = 0$
    $x = -3, 4$

16. $x^2 - 14x + 49 = 0$
    $(x-7)^2 = 0$
    $x = 7$

17. $\frac{1}{2}(x - 2)^2 = 2$
     $(x - 2)^2 = 4$
     $(x - 2) = \pm 2$
     $x = 0, 4$

18. $6x - 3 = 5 + x^2$
     $x^2 - 6x + 8 = 0$
     $(x-4)(x-2) = 0$
     $x = 2, 4$

19. $(3x - 5)^2 - 5 = 0$
     $(3x - 5)^2 = 5$
     $3x - 5 = \pm \sqrt{5}$
     $3x = 5 \pm \sqrt{5}$
     $x = 1/3(5 \pm \sqrt{5})$

20. $\sqrt{2x+5} = \sqrt{x^2+2}$
     $2x + 5 = x^2 + 2$
     $x^2 - 2x - 3 = 0$
     $(x+1)(x-3) = 0$
     $x = -1, 3$

◆◆◆◆◆

7.2 SOLVING ANY QUADRATIC EQUATIONS

OBJECTIVES:

*(A) To solve a quadratic equation by completing the square.

** REVIEW

In order to complete the square, consider the following:

1. The coefficient of the x^2 term must be 1.

2. $x^2 + bx + (b/2)^2 = [x + (b/2)]^2$ --> the last term must be the square of half the coefficient of the x term.

Examples: Complete the squares of the following:

a. $x^2 - 8x$ b. $x^2 + x$ c. $x^2 + \frac{1}{4}$

Solutions:
    a. $x^2 - 8x$
       $x^2 - 8x + [\frac{1}{2}(8)]^2$ --> last term $[\frac{1}{2}(8)]^2$ is the square of
                                    half the coefficient of the x term
    b. $x^2 + x$
       $x^2 + x + [\frac{1}{2}(1)]^2$ --> last term $[\frac{1}{2}(1)]^2$ is the square of
                                    half the coefficient of the x term
    c. $x^2 + \frac{1}{4}x$
       $x^2 + \frac{1}{4}x + [\frac{1}{2}(\frac{1}{4})]^2$ --> last term $[\frac{1}{2}(\frac{1}{4})]^2$ is the square of
                                    half the coefficient of the x term

● To solve a quadratic equation by completing the square remember the strategy for completing the square discussed in Elementary Algebra by Mark Dugopolski.

Example: Solve by completing the square.

    $3x^2 + 5x + 2 = 0$

    $x^2 + 5/3x + 2/3 = 0$ --> the coefficient of x^2 term is 1

160

The square of half the coefficient of the x term is $[\frac{1}{2}(5/3)]^2$

Add $[\frac{1}{2}(5/3)]^2) = (25/36)$ to both sides of the equation.

$(x^2 + 5/3x + 25/36) + 2/3 = 25/36$ --- simplify

$(x + 5/6)^2 = 25/36 - 2/3$

$(x + 5/6)^2 = 1/36$ --- solve the simple quadratic equation

$(x + 5/6) = \pm 1/6$

$x = -1$ or $x = -2/3$.

◆◆◆◆◆

**EXERCISES

I. Factor each perfect trinomial as the square of a binomial.

1. $x^2 - 4x + 4$ 2. $x^2 - \frac{1}{2}x + 1/16$

2. $x^2 - 1/3x + 1/36$ 4. $x^2 + 3/5x + 9/100$

II. Solve each quadratic equation by completing the square.

5. $x^2 - 2x + 19 = 0$ 6. $3x^2 - 8x + 7/3 = 0$

7. $3t^2 + 4t + 1 = 0$ 8. $3x^2 + 7x + 1 = 0$

9. $4x^2 - 4x + 1 = 0$ 10. $2x^2 + x - 14 = 0$

11. $x^2 + 6x + 3 = 0$ 12. $x^2 + x + 1 = 0$

III. Solve each equation.

13. $x^2 - 12 = 0$ 14. $y^2 + 3y = 0$

15. $x^2 - 8x - 9 = 0$ 16. $(x+3)^2 = 1$

17. $3x^2 + 9x = 1$ 18. $(7x+1)^2 = 0$

19. $5x^2 + 10x = -5$ 20. $3y^2 - 7y + 2 = 0$

◆◆◆◆◆

**SOLUTIONS TO EXERCISES

I. Factor each perfect trinomial as the square of a binomial.

1. $x^2 - 4x + 4$ 2. $x^2 - \frac{1}{2}x + 1/16$
   $= (x - 2)^2$ $= (x - \frac{1}{4})^2$

2. $x^2 - 1/3x + 1/36$ 4. $x^2 + 3/5x + 9/100$
   $= (x - 1/6)^2$ $= (x + 3/10)^2$

II. Solve each quadratic equation by completing the square.

5. $x^2 - 2x + 19 = 0$
   $(x^2 - 2x + 1) + 18 = 0$
   $(x^2 - 2x + 1) = -18$
   $(x - 1)^2 = -18$
   $x - 1 = \pm 3\sqrt{2}i$
   $x = 1 \pm 3\sqrt{2}i$

6. $3x^2 - 8x + 7/3 = 0$
   $x^2 - (8/3)x + 7/9 = 0$
   $x^2 - (8/3)x + 16/9 = -(7/9) + (16/9)$
   $(x - 4/3)^2 = 1$
   $x - 4/3 = \pm 1$
   $x = 4/3 \pm 1$

7. $3t^2 + 4t + 1 = 0$
   $t^2 + 4/3t + 1/3 = 0$
   $(t + 2/3)^2 = 1/9$
   $t = -1/3, -1$

8. $3x^2 + 7x + 1 = 0$
   $x^2 + (7/3)x + 1/3 = 0$
   $(x + 7/6)^2 = 37/36$
   $x = 1/6(\pm\sqrt{37} - 7)$

9. $4x^2 - 4x + 1 = 0$
   $(2x - 1)^2 = 0$
   $x = \tfrac{1}{2}$

10. $2x^2 + x - 14 = 0$
    $(x + \tfrac{1}{4})^2 = 7 + (\tfrac{1}{4})^2$
    $x + \tfrac{1}{4} = \pm \tfrac{1}{4}(\sqrt{113})$
    $x = \pm \tfrac{1}{4}(\sqrt{113}) - \tfrac{1}{4}$

11. $x^2 + 6x + 3 = 0$
    $(x + 3)^2 = 6$
    $x + 3 = \pm \sqrt{6}$
    $x = \pm \sqrt{6} - 3$

12. $x^2 + x + 1 = 0$
    $(x + \tfrac{1}{2})^2 = -3/4$
    $x + \tfrac{1}{2} = \pm \tfrac{1}{2}(\sqrt{3})i$
    $x = \tfrac{1}{2}[\pm(\sqrt{3})i - 1]$

III. Solve each equation.

13. $x^2 - 12 = 0$

    $x^2 = 12$

    $x = \pm 2\sqrt{3}$

14. $y^2 + 3y = 0$

    $y(y + 3) = 0$

    $y = 0, -3$

15. $x^2 - 8x - 9 = 0$
    $(x + 1)(x - 9) = 0$

    $x = -1, 9$

16. $(x+3)^2 = 1$
    $x + 3 = \pm 1$

    $x = -4, -2$

17. $3x^2 + 9x = 1$

    $x^2 + 3x = 1/3$

    $x^2 + 3x + [3/2]^2 = 1/3 + (3/2)^2$

    $(x + 3/2)^2 = 31/12$

    $x = \pm \sqrt{31}/2\sqrt{3} - 3/2$

18. $(7x + 1)^2 = 0$

    $7x + 1 = 0$

    $x = -1/7$

19. $5x^2 + 10x = -5$

    $5(x + 1)^2 = 0$

    $x = -1$

20. $3y^2 - 7y + 2 = 0$

    $(3y - 1)(y - 2) = 0$

    $y = 1/3, 2$

♦♦♦♦♦

162

7.3 THE QUADRATIC FORMULA

OBJECTIVES:

*(A) To solve any quadratic equation by using the quadratic formula.

** REVIEW

*(A) To solve a quadratic equation $ax^2 + bx + c = 0$ by using the quadratic formula.

$$x = \frac{-b \pm \sqrt{b^2 - 4ac}}{2a}$$

- Where a is the coefficient of x^2 term, b is the coefficient of x term, and c is the constant.

- To solve a quadratic equation by using the formula you should write the given equation in the form of $(ax^2 + bx + c) = 0$. Then identify the values of a,b and c and substitute them into the quadratic formula.

- The quantity $(b^2 - 4ac)$ is called the discriminant of the quadratic equation.

- The discriminant determines the type of solutions of $ax^2 + bx + c = 0$.

$(b^2-4ac) > 0 \longrightarrow$ **two real roots**

$(b^2-4ac) = 0 \longrightarrow$ **one real root**

$(b^2-4ac) < 0 \longrightarrow$ **two complex roots**

Examples:
   a. $x^2 + 3x - 10 = 0$ b. $x^2 - 6x + 9 = 0$

Solutions:

a. $x^2 + 3x - 10 = 0$ b. $x^2 - 6x + 9 = 0$
   a=1, b=3, and c=-10   a=1, b=-6, and c=9
       (Substitute these values into the quadratic formula)

$x = \dfrac{-3 \pm \sqrt{3^2 - 4(1)(-10)}}{2(1)}$ $x = \dfrac{-(-6) \pm \sqrt{(-6)^2 - 4(1)(9)}}{2(1)}$

$x = \dfrac{-3 \pm \sqrt{9 + 40}}{2}$ $x = \dfrac{6 \pm \sqrt{0}}{2}$

$x = \dfrac{-3 \pm \sqrt{49}}{2}$ $x = \dfrac{6}{2}$

$x = 2, -5$ $x = 3$

163

c. $x^2 + 4x + 5 = 0$ where a=1, b=4, and c=5.

$$x = \frac{-4 \pm \sqrt{(4)^2 - 4(1)(5)}}{2(1)}$$

$$x = \frac{-4 \pm \sqrt{16 - 20}}{2}$$

$$x = \frac{-4 \pm \sqrt{-4}}{2}$$

$$x = -2 \pm i$$

● Methods of solving quadratic equations :

1. Solve by using the square root property when the quadratic equation is of the form $x^2 = c$. Then $x = \pm\sqrt{c}$.
2. Solve by factoring the equation when the equation can be written as $(ax - b)(cx - d) = 0$.
3. Solve by using the quadratic formula.

◆◆◆◆◆

**EXERCISES

I. Solve by the quadratic formula.

1. $x^2 + 3x + 2 = 0$ 2. $4x^2 - 12x + 9 = 0$
3. $x^2 - 2x - 8 = 0$ 4. $x^2 + 6x + 8 = 0$

II. Find the value of the discriminant and state how many real solutions there are to each quadratic equation.

5. $3x^2 + 6x + 3 = 0$ 6. $2x^2 + 3x + 4 = 0$
7. $-5x^2 + 9x + 1 = 0$ 8. $3x^2 + 3x + 1 = 0$
9. $x - 3 = 4x^2$ 10. $y^2 - y - 1 = 0$

III. Solve each of the following equations.

11. $x^2 + 10x + 9 = 0$ 12. $x^2 - 4x = 0$
13. $x^2 + 3x + 4 = 0$ 14. $2x^2 - 5x + 2 = 0$
15. $x^2 - 4x + 1 = 0$ 16. $x^2 - x - 1 = 0$
17. $x^2 = 3$ 18. $(x+1)(x-2) = (2x+1)^2$

19. $\dfrac{x}{2} = \dfrac{-1}{x+2}$ 20. $\dfrac{x-3}{8} = \dfrac{x}{x+3}$

◆◆◆◆◆

**SOLUTIONS TO EXERCISES

I. Solve by the quadratic formula.
 Hint:
(Identify the values of a, b, and c then substitute into the formula.)

1. $x^2+3x+2=0$

   $x=1, b=3, and\ c=2$

   $x = \dfrac{-3\pm\sqrt{3^2-4(1)(2)}}{2(1)}$

   $x = \dfrac{-3\pm1}{2}$

   $x = -2, -1$

2. $4x^2-12x+9=0$

   $a=4, b=-12, and\ c=9$

   $x = \dfrac{-(-12)\pm\sqrt{(-12)^2-4(4)(9)}}{2(4)}$

   $x = \dfrac{12\pm0}{8}$

   $x = \dfrac{3}{2}$

3. $x^2-2x-8=0$

   $a=1, b=-2, and\ c=-8$

   $x = \dfrac{-(-2)\pm\sqrt{(-2)^2-4(1)(-8)}}{2(1)}$

   $x = \dfrac{2\pm6}{2}$

   $x = 4, -2$

4. $x^2+6x+8=0$

   $a=1, b=6, and\ c=8$

   $x = \dfrac{(-6)\pm\sqrt{(6)^2-4(1)(8)}}{2(1)}$

   $x = \dfrac{-6\pm2}{2}$

   $x = -2, -4$

II. Find the value of the discriminant and state how many real solutions there are to each quadratic equation. The discriminant is the quantity $(b^2 - 4ac)$.

5. $3x^2 + 6x + 3 = 0$
   $(6)^2 - 4(3)(3) = 0$
   One real root.

6. $2x^2 + 3x + 4 = 0$
   $(3)^2 - 4(2)(4) = -23$
   Two complex roots.

7. $-5x^2 + 9x + 1 = 0$
   $(9)^2 - 4(-5)(1) = 101 > 0$
   Two real roots.

8. $3x^2 + 3x + 1 = 0$
   $(3)^2 - 4(3)(1) < 0$
   Two complex roots.

9. $x - 3 = 4x^2$
   $4x^2 - x - 3 = 0$
   $(-1)^2 - 4(4)(3) < 0$
   Two complex roots.

10. $y^2 - y - 1 = 0$
    $a=1, b=-1, and\ c=-1$
    $(-1)^2 - 4(1)(-1) > 0$
    Two real roots.

III. Solve each equation.

11. $x^2 + 10x + 9 = 0$
    $(x + 9)(x + 1) = 0$
    $x = -9, -1$

12. $x^2 - 4x = 0$
    $x(x - 4) = 0$
    $x = 0, 4$

13. $x^2 + 3x + 4 = 0$
    $a=1, b=3, and\ c=4$
    $x = \frac{1}{2}(-3\pm\sqrt{7}i)$

14. $2x^2 - 5x + 2 = 0$
    $a=2, b=-5, and\ c=2$
    $x = \frac{1}{2}, 2$

15. $x^2 - 4x + 1 = 0$
    $a=1, b=-4, and\ c=1$
    $x = 2 \pm \sqrt{3}$

16. $x^2 - x - 1 = 0$
    $a=1, b=-1, and\ c=-1$
    $x = \frac{1}{2}(1 \pm \sqrt{5})$

17. $x^2 = 3$

$x = \pm\sqrt{3}$

18. $(x+1)(x-2) = (2x+1)^2$

$x^2 - x - 2 = 4x^2 + 4x + 1$

$3x^2 + 5x + 3 = 0$

$a=3, \ b=5, \ and \ c=3$

$x = 1/6(-5\pm\sqrt{11}i)$

19. $\dfrac{x}{2} = \dfrac{-1}{x+2}$

$x(x+2) = -2$

$x^2 + 2x + 2 = 0$

$x = \dfrac{(-2)\pm\sqrt{4-8}}{2}$

$x = \dfrac{-2\pm2i}{2}$

$x = -1\pm i$

$(x \neq -2)$

20. $\dfrac{x-3}{8} = \dfrac{x}{x+3}$

$x^2 - 9 = 8x$

$x^2 - 8x - 9 = 0$

$x = \dfrac{8\pm\sqrt{64+36}}{2}$

$x = 4\pm5$

$x = 9, \ -1$

$(x \neq -3)$

♦♦♦♦♦

7.5 COMPLEX NUMBERS

OBJECTIVES:

*(A) To perform arithmetic operations with complex numbers.
*(B) To solve quadratic equations with complex solutions.

** REVIEW

*(A) A complex number is a number that has the form (a + bi), where a and b are real numbers ($i = \sqrt{-1}$ and $i^2 = -1$).

These are examples of complex numbers

$3\sqrt{-1} = 3i$, $\sqrt{-5} = \sqrt{5}\sqrt{-1} = \sqrt{5}i$, and $3 - 5i$

- A complex number (**a** + **bi**) has two parts:
  a is called the real part and **bi** is the imaginary part.

Examples:
  3 - 5i --→ The real part is 3 and the imaginary part is -5i

  $\sqrt{5}i$ ---→ The real part is 0 and the imaginary part is $\sqrt{5}i$

- Two complex numbers (a + bi) and (c + di) are equal if and only if a = c and b = d.
  (The real parts are equal to each other and the imaginary parts are equal to each other).

- The two complex numbers **a + bi** and **a - bi** are called conjugate of each other. Their product is always a real number.

166

$(a + bi) \cdot (a - bi) = (a)^2 - (bi)^2 = a^2 - b^2 i^2 = a^2 + b^2$.

Algebraic operations with complex numbers:

1. Addition. $(a + bi) + (c + di) = (a + c) + (b + d)i$

2. Subtraction. $(a + bi) - (c + di) = (a - c) + (b - d)i$

3. Multiplication. $(a + bi) \cdot (c + di) = (ac - bd) + (ad + bc)i$
   use the FOIL method to multiply and remember $i^2 = -1$.

4. *Division.* $\dfrac{a + bi}{c + di} = \dfrac{(ac+bd) + (bc-ad)i}{c^2 + d^2}$

*(B) To solve a quadratic equation with complex solutions, do the following:

1. If it is a simple quadratic equation $x^2 = c$ ($c \leq 0$), then $x = \pm\sqrt{c}\,i$ is the solution.

2. If $ax^2 + bx + c = 0$ ($a \neq 0$ and $b \neq 0$), then apply the quadratic formula to solve the equation.

Examples:
a. $x^2 = -81$ b. $x^2 - 4x + 13 = 0$

Solutions:
a. $x^2 = -81$ This is a simple quadratic equation.

$x = \pm\sqrt{81}\,i$

$x = \pm 9i$

b. $x^2 - 4x + 13 = 0$

$a = 1$, $b = -4$, *and* $c = 13$

$x = \dfrac{-(-4) \pm \sqrt{(-4)^2 - 4(1)(13)}}{2}$

$x = \dfrac{-(-4) \pm \sqrt{16 - 52}}{2} = \dfrac{4 \pm \sqrt{-36}}{2} = \dfrac{4 \pm 6i}{2}$

$x = 2 + 3i,\ 2 - 3i$

One solution is the conjugate of the other.

◆◆◆◆◆

167

◆◆◆◆◆

**EXERCISES

I. Perform the indicated operations.

1. $(-3 + 5i) + 6$

2. $(2 - 5i) - (3 + 8i)$

3. $(8 - i)(8 + i)$

4. $(2i + 1)^2$

5. $(2i + 3)(3 - 2i)$

6. $(-2i)^3$

7. $(4 + i)(-i)$

8. $(12i + 10) \div (2i)$

9. $(3 + 5i)(3 - 5i)$

10. $(2 + 3i)(3 + 2i)$

II. Find the complex solutions to each quadratic equation.

11. $x^2 + 16 = 0$

12. $x^2 + 1 = 0$

13. $3x^2 + 12 = 0$

14. $3x^2 + 147 = 0$

15. $x^2 + 2x + 3 = 0$

16. $x^2 + 2x + 9 = 0$

17. $x^2 - x + 1 = 0$

18. $7x^2 + x + 1 = 0$

19. What is the value of $x^2 - x - 1$, if $x = 2 + i$

20. What is the value of $3x^2 - x + 5$, if $x = i - 3$

◆◆◆◆◆

**SOLUTIONS TO EXERCISES

I. Perform the indicated operations.

1. $(-3 + 5i) + 6$
   $= 3 + 5i$

2. $(2 - 5i) - (3 + 8i)$
   $= -(1 + 13i)$

3. $(8 - i)(8 + i)$
   $= (8)^2 - i^2 = 64 + 1 = 65$

4. $(2i + 1)^2$
   $= 4i - 3$

5. $(2i + 3)(3 - 2i)$
   $= (3+2i)(3-2i)=(3)^2 -4(i)^2$
   $= 9 + 4 = 13$

6. $(-2i)^3$
   $= -8i^3 = -8i(i^2)$
   $= 8i$

7. $(4 + i)(-i)$
   $= -4i - i^2$
   $= 1 - 4i$

8. $(12i + 10) \div (2i)$
   $= (12i)/(2i)+ (10/2i) = 6 + 5/i$
   $= 6 + (5/i)\cdot(i/i)= 6 - 5i$

9. $(3 + 5i)(3 - 5i)$
   $= 9 - 25i^2$
   $= 9 - 25(-1)$
   $= 34$

10. $(2 + 3i)(3 + 2i)$
   $= 6 + 4i + 9i + 6i^2$
   $= 13i + 6 - 6$
   $= 13i$

II. Find the complex solutions to each quadratic equation.

11. $x^2 + 16 = 0$
   $x^2 = -16$
   $x = \pm 4i$

12. $x^2 + 1 = 0$
   $x^2 = -1$
   $x = \pm i$

13. $3x^2 + 12 = 0$
   $x^2 = -4$
   $x = \pm 2i$

14. $3x^2 + 147 = 0$
   $x^2 = -49$
   $x = \pm 7i$

15. $x^2 + 2x + 3 = 0$
   $a = 1, b = 2, $ and $c = 3$

   $x = \dfrac{-2 \pm \sqrt{2^2 - 4(1)(3)}}{2(1)}$

   $x = \dfrac{-2 \pm \sqrt{4 - 12}}{2} = -1 \pm \sqrt{2}\,i$

16. $x^2 + 2x + 9 = 0$
   $a = 1, b = 2, $ and $c = 9$

   $x = \dfrac{(-2) \pm \sqrt{(2)^2 - 4(1)(9)}}{2(1)}$

   $x = \dfrac{(-2) \pm \sqrt{32}\,i}{2} = -1 \pm 2\sqrt{2}\,i$

17. $x^2 = x + 1 = 0$
   $a = 1, b = -1, $ and $c = 1$

   $x = \dfrac{-(-1) \pm \sqrt{(-1)^2 - 4(1)(1)}}{2(1)}$

   $x = \dfrac{1 \pm \sqrt{1 - 4}}{2} = \dfrac{1 \pm \sqrt{3}\,i}{2}$

18. $7x^2 + x + 1 = 0$
   $a = 7, b = 1, $ and $c = 1$

   $x = \dfrac{(-1) \pm \sqrt{(1)^2 - 4(7)(1)}}{2(7)}$

   $x = \dfrac{(-1) \pm \sqrt{1 - 28}}{14} = \dfrac{(-1) \pm 3\sqrt{3}\,i}{14}$

19. What is the value of $x^2 - x - 1$, if $x = 2 + i$

   $x^2 - x - 1$
   $= (2 + i)^2 - (2 + i) - 1$
   $= 4 + 4i + i^2 - 2 - i - 1$
   $= 3i$

20. What is the value of $3x^2 - x + 5$, if $x = i - 3$

   $3x^2 - x + 5$
   $= 3(i - 3)^2 - (i - 3) + 5$
   $= 3i^2 - 18i + 27 - i + 3 + 5$
   $= 32 - 19i$

CHAPTER VII - TEST A

I. Solve each equation.

    1. $x^2 - 16 = 0$ 2. $x^2 + 16 = 0$ 3. $x^2 = 5$

    4. $(x - 3)^2 = 100$ 5. $2x^2 + 5x + 2 = 0$ 6. $x^2 + x = 0$

    7. $16x^2 + 8x + 1 = 0$ 8. $x^2 - x - 1 = 0$ 9. $(x + 5)^2 = 1$

II. Solve by completing the square.

    10. $x^2 - 2x + 5 = 0$ 11. $9x^2 + 1 = -6x$

    12. $5m^2 + 25 = 0$ 13. $3x^2 - x = -1$

III. Solve each equation.

    14. $2x(x + 5) = 1$ 15. $t^2 - 3t + 4 = 0$

    16. $x^2 - 6x + 5 = 0$ 17. $3x^2 - 11x + 10 = 0$

    18. $x^2 = 49 + 14x$ 19. $(x - 2)(x + 5) = -12$

    20. $x^2 - x + \frac{1}{4} = 0$

♦♦♦♦♦

CHAPTER VII - TEST B

I. Use the discriminant to determine whether the following equations
have solutions that are (a) two real roots, (b) one real root,
(c) two complex roots.

    1. $x^2 + 3x + 5 = 0$ 2. $-3x^2 + + 9x + 1 = 0$

    3. $2y^2 - 4y + 2 = 0$ 4. $3x^2 = 2 + 5x$

    5. $3m^2 + 6m + 10 = 0$ 6. $x^2 + 3x + 3 = 0$

    7. $4x^2 - 20x + 11 = 0$ 8. $3x^2 + 9 = 0$

    9. $9x^2 = x + 1$ 10. $4x^2 - 4x + 1 = 0$

II. Solve each equation.

    11. $x^2 + 25 = 0$ 12. $x^2 - 25 = 0$

    13. $x^2 - 11x - 10 = 0$ 14. $x^2 + x - 3 = 0$

    15. $y^2 - 2y + 4 = 0$ 16. $3x^2 + 7x + 1 = 0$

17. $x^2 + 2x + 2 = 0$ 18. $y^2 + 6y + 6 = 0$

19. $4x^2 + 4x - 1 = 0$ 20. $x^2 - 2x = 4$

◆◆◆◆◆

CHAPTER VII - TEST C

I. Perform the indicated operations. Write the answer in the form **a + bi**

1. $(-3 + 8i) + (2 + i) + 5i$ 2. $(1 + 3i)(i - 3)$

3. $(6 - 5i) - (3 + 8i)$ 4. $4i - 2(7 - 3i)$

5. $(3 - i)(-3 + i)$ 6. $(2i + 1)^2$

7. $(1 + i)^2 + (5i)^2$ 8. $i^2 + 2i + 1$

9. $\dfrac{5 - \sqrt{-50}}{5}$ 10. $\dfrac{3 - 2i}{i}$

11. $\dfrac{i}{i - 2}$ 12. $\dfrac{1 - 3i}{3i + 2}$

II. Find the complex solutions to the quadratic equations.

13. $4x^2 + 1 = 0$ 14. $x^2 + 7 = 0$

15. $x^2 + 10x + 100 = 0$ 16. $3x^2 - x + 1 = 0$

17. $x^2 - 3x + 10 = 0$ 18. $9x^2 - x + 1 = 0$

19. $2x^2 - 3x + 5 = 0$ 20. $x^2 =- x + 5 = 0$

◆◆◆◆◆

CHAPTER VII - TEST A - SOLUTIONS

I. Solve each equation.

1. $x^2 - 16 = 0$
   $x = \pm 4$

2. $x^2 + 16 = 0$
   $x = \pm 4i$

3. $x^2 = 5$
   $x = \pm\sqrt{5}$

4. $(x - 3)^2 = 100$
   $x = \pm 10 + 3$
   $x = -7, 13$

5. $2x^2 + 5x + 2 = 0$
   $(x + 2)(2x + 1) = 0$
   $x = -\frac{1}{2}, -2$

6. $x^2 + x = 0$
   $x(x + 1) = 0$
   $x = 0, -1$

7. $16x^2 + 8x + 1 = 0$
   $(4x + 1)^2 = 0$
   $x = -1/4$

8. $x^2 - x - 1 = 0$
   $x = \frac{1}{2}[1 \pm \sqrt{5}]$

9. $x + 5)^2 = 1$
   $x + 5 = \pm 1$
   $x = -6, -4$

II. Solve by completing the square.

10. $x^2 - 2x + 5 = 0$
    $x^2 - 2x + 1 = -5 + 1$
    $(x - 1)^2 = -4$
    $x - 1 = \pm 2i$
    $x = 1 \pm 2i$

11. $9x^2 + 1 = -6x$
    $9x^2 + 6x + 1 = 0$
    $(3x + 1)^2 = 0$
    $x = -1/3$

12. $5m^2 + 25 = 0$
    $5(m^2 + 5) = 0$
    $m^2 + 5 = 0$
    $m^2 = -5$
    $m = \pm\sqrt{5}i$

13. $3x^2 - x + 1 = 0$
    $3x^2 - x = -1$
    $x^2 - (1/3)x + 1/36 = -1/3 + (1/36)$
    $(x - 1/6)^2 = -11/36$
    $x = 1/6[1 \pm\sqrt{11}i]$

III. Solve each equation.

14. $2x(x + 5) = 1$
    $2x^2 + 10x - 1 = 0$

    $a=2, b=10,$ and $c=-1$

    $x = \dfrac{-10 \pm \sqrt{(10)^2 - 4(2)(-1)}}{2(2)}$

    $x = \dfrac{(-10) \pm \sqrt{100+8}}{4} = \dfrac{-5 \pm 3\sqrt{3}}{2}$

15. $t^2 - 3t + 4 = 0$

    $a=1, b=-3$ and $c=4$

    $t = \dfrac{-(-3) \pm \sqrt{(-3)^2 - 4(1)(4)}}{2(1)}$

    $t = \dfrac{3 \pm \sqrt{7}\,i}{2}$

16. $x^2 - 6x + 5 = 0$
    $(x - 1)(x - 5) = 0$
    $x = 1, 5$

17. $3x^2 - 11x + 10 = 0$
    $(3x - 5)(x - 2) = 0$
    $x = 5/3, 2$

18. $x^2 = 49 + 14x$
    $x^2 - 14x - 49 = 0$
    $(x - 7)^2 = 0$
    $x = 7$

19. $(x - 2)(x + 5) = -12$
    $x^2 + 3x - 10 = -12$
    $x^2 + 3x + 2 = 0$
    $(x + 2)(x + 1) = 0$
    $x = -2, -1$

20. $x^2 - x + \frac{1}{4} = 0$
    $(x - \frac{1}{2})^2 = 0$
    $x = \frac{1}{2}$

♦♦♦♦♦

CHAPTER VII - TEST B - SOLUTIONS

I. Use the discriminant to determine whether the following equations have solutions that are (a) two real roots, (b) one real root, (c) two complex roots.

- The discriminant is the quantity $(b^2 - 4ac)$

1. $x^2 + 3x + 5 = 0$
$a = 1, b = 3,$ and $c = 5$
$3^2 - 4(1)(5) = -11 < 0$
Two complex roots.

2. $-3x^2 + 9x + 1 = 0$
$a = -3, b = 9,$ and $c = 1$
$9^2 - 4(-3)(1) = 93 > 0$
Two real roots.

3. $2y^2 - 4y + 2 = 0$
$b^2 - 4ac = 0$
One real root.

4. $3x^2 = 2 + 5x$
$b^2 - 4ac = 49 > 0$
Two real roots.

5. $3m^2 + 6m + 10 = 0$
$b^2 - 4ac = -84 < 0$
Two complex roots.

6. $x^2 + 3x + 3 = 0$
$b^2 - 4ac = -3 < 0$
Two complex roots.

7. $4x^2 - 20x + 11 = 0$
$b^2 - 4ac = 224 > 0$
Two real roots.

8. $3x^2 + 9 = 0$
$b^2 - 4ac = -108 < 0$
Two complex roots

9. $9x^2 = x + 1$
$b^2 - 4ac = 37 > 0$
Two real roots.

10. $4x^2 - 4x + 1 = 0$
$b^2 - 4ac = 0$
One real root.

II. Solve each equation.

11. $x^2 + 25 = 0$
$x^2 = -25$
$x = \pm 5i$

12. $x^2 - 25 = 0$
$x^2 = 25$
$x = \pm 5$

13. $x^2 - 11x - 10 = 0$
$a = 1, b = -11,$ and $c = -10$

$$x = \frac{-(-11) \pm \sqrt{(-11)^2 - 4(1)(-10)}}{2(1)}$$

$$x = \frac{11 \pm \sqrt{121 + 40}}{2}$$

$$x = \frac{11 \pm \sqrt{161}}{2}$$

14. $x^2 + x - 3 = 0$
$a = 1, b = 1,$ and $c = -3$

$$x = \frac{(-1) \pm \sqrt{(1)^2 - 4(1)(-3)}}{2(1)}$$

$$x = \frac{(-1) \pm \sqrt{1 + 12}}{2}$$

$$x = \frac{(-1) \pm \sqrt{13}}{2}$$

15. $y^2-2y+4=0$

$a=1$, $b=-2$, and $c=4$

$$x = \frac{-(-2) \pm \sqrt{(-2)^2-4(1)(4)}}{2(1)}$$

$$x = \frac{2 \pm \sqrt{4-16}}{2}$$

$$x = \frac{2 \pm 2\sqrt{3}\,i}{2}$$

$$x = 1 \pm \sqrt{3}\,i$$

16. $3x^2+7x+1=0$

$a=3$, $b=7$, and $c=1$

$$x = \frac{(-7) \pm \sqrt{(7)^2-4(3)(1)}}{2(3)}$$

$$x = \frac{(-7) \pm \sqrt{49-12}}{6}$$

$$x = \frac{-7 \pm \sqrt{37}}{6}$$

17. $x^2+2x+2=0$

$a=1$, $b=2$, and $c=2$

$$x = \frac{(-2) \pm \sqrt{(2)^2-4(1)(2)}}{2(1)}$$

$$x = \frac{(-2) \pm \sqrt{4-8}}{2}$$

$$x = \frac{-2 \pm 2i}{2}$$

$$x = -1 \pm i$$

18. $x^2+6x+6=0$

$a=1$, $b=6$, and $c=6$

$$x = \frac{(-6) \pm \sqrt{(6)^2-4(1)(6)}}{2(1)}$$

$$x = \frac{(-6) \pm \sqrt{36-24}}{2}$$

$$x = \frac{(-6) \pm 2\sqrt{3}}{2}$$

$$x = -3 \pm \sqrt{3}$$

19. $4x^2+4x-1=0$

$a=4$, $b=4$, and $c=-1$

$$x = \frac{(-4) \pm \sqrt{(4)^2-4(-1)(4)}}{2(4)}$$

$$x = \frac{-4 \pm \sqrt{32}}{8}$$

$$x = \frac{-4 \pm 4\sqrt{2}}{8}$$

$$x = \frac{-1 \pm \sqrt{2}}{2}$$

20. $x^2-2x-4=0$

$a=1$, $b=-2$, and $c=-4$

$$x = \frac{-(-2) \pm \sqrt{(-2)^2-4(1)(-4)}}{2(1)}$$

$$x = \frac{(2) \pm \sqrt{4+16}}{2}$$

$$x = \frac{2 \pm \sqrt{20}}{2}$$

$$x = 1 \pm \sqrt{5}$$

CHAPTER VII - TEST C - SOLUTIONS

I. Perform the indicated operations.
Write the answer in the form **a + bi**.

1. $(-3 + 8i) + (2 + i) + 5i$
   $= (-3 + 2) + (8i + i + 5i)$
   $= -1 + 14i$

2. $(1 + 3i)(i - 3)$
   $= (i - 3 + 3i^2 - 9i)$
   $= -6 - 8i$

3. $(6 - 5i) - (3 + 8i)$
   $= (6 - 3) + (-5 - 8)i$
   $= 3 - 13i$

4. $4i - 2(7 - 3i)$
   $= 4i - 14 + 6i$
   $= -14 + 10i$

5. $(3 - i)(-3 + i)$
   $= -9 + 3i + 3i - i^2$
   $= -8 + 6i$

6. $(2i + 1)^2$
   $= (2i)^2 + 4i + 1$
   $= 4i^2 + 4i + 1$
   $= 4i - 3$

7. $(1 + i)^2 + (5i)^2$
   $= 1 + 2i + i^2 + 25i^2$
   $= -25 + 2i$

8. $i^2 + 2i + 1$
   $= (-1) + 2i + 1$
   $= 2i$

9. $\dfrac{5 - \sqrt{-50}}{5}$

   $= \dfrac{5 - 5\sqrt{2}\,i}{5}$

   $= 1 - \sqrt{2}\,i$

10. $\dfrac{3 - 2i}{i}$

    $= \dfrac{3 - 2i}{i} \cdot \dfrac{i}{i}$

    $= -(2 + 3i)$

11. $\dfrac{i}{i - 2}$

    $= \left(\dfrac{i}{i - 2}\right) \cdot \dfrac{(i + 2)}{(i + 2)}$

    $= \dfrac{i^2 + 2i}{i^2 - 4}$

    $= \dfrac{(1 - 2i)}{5}$

12. $\dfrac{1 - 3i}{3i + 2}$

    $= \left(\dfrac{1 - 3i}{3i + 2}\right) \cdot \dfrac{(3i - 2)}{(3i - 2)}$

    $= \dfrac{(3i - 2 - 9i^2 + 6i)}{(9i^2 - 4)}$

    $= -\dfrac{7 + 9i}{13}$

II. Find the complex solutions to the quadratic equations.

13. $4x^2 + 1 = 0$
    $4x^2 = -1$
    $x = \pm \tfrac{1}{2}i$

14. $x^2 + 7 = 0$
    $x^2 = -7$
    $x = \pm\sqrt{7}\,i$

15. $x^2+10x+100=0$

$a=1$, $b=10$, and $c=100$

$$x = \frac{(-10) \pm \sqrt{(10)^2 - 4(1)(100)}}{2(1)}$$

$$x = \frac{-10 \pm \sqrt{100-400}}{2}$$

$$x = -5 \pm 5\sqrt{3}\,i$$

16. $3x^2-x+1=0$

$a=3$, $b=-1$, and $c=1$

$$x = \frac{-(-1) \pm \sqrt{(-1)^2 - 4(1)(3)}}{2(3)}$$

$$x = \frac{(1) \pm \sqrt{1-12}}{6}$$

$$x = \frac{(1) \pm \sqrt{11}\,i}{6}$$

17. $x^2-3x+10=0$

$a=1$, $b=-3$, and $c=10$

$$x = \frac{-(-3) \pm \sqrt{(-3)^2 - 4(1)(10)}}{2(1)}$$

$$x = \frac{3 \pm \sqrt{9-40}}{2}$$

$$x = \frac{3 \pm \sqrt{31}\,i}{2}$$

18. $9x^2-x+1=0$

$a=9$, $b=-1$, and $c=1$

$$x = \frac{-(-1) \pm \sqrt{(-1)^2 - 4(9)(1)}}{2(9)}$$

$$x = \frac{(1) \pm \sqrt{1-36}}{18}$$

$$x = \frac{1 \pm \sqrt{35}\,i}{18}$$

19. $2x^2-3x+5=0$

$a=2$, $b=-3$, and $c=5$

$$x = \frac{-(-3) \pm \sqrt{(-3)^2 - 4(2)(5)}}{2(2)}$$

$$x = \frac{3 \pm \sqrt{9-40}}{4}$$

$$x = \frac{3 \pm \sqrt{31}\,i}{4}$$

20. $x^2-x+5=0$

$a=1$, $b=-1$, and $c=5$

$$x = \frac{-(-1) \pm \sqrt{(-1)^2 - 4(1)(5)}}{2(1)}$$

$$x = \frac{(1) \pm \sqrt{1-20}}{2}$$

$$x = \frac{1 \pm \sqrt{19}\,i}{2}$$

◆◆◆◆◆

CHAPTER 8

LINEAR EQUATIONS IN TWO VARIABLES

8.1 THE RECTANGULAR COORDINATE SYSTEM

OBJECTIVES:

*(A) To understand the ordered pair and the rectangular
     coordinate system.
*(B) To plot points (ordered pairs) and graph lines.

** REVIEW

*(A) Consider the equation y = 3x - 5.
  The following pairs of x values and y values satisfy the equation.
  (0, -5) -- substitute x=0 into the equation and find y=3(0)-5=-5.
  (1, -2) -- substitute x=1 into the equation and find y=3(1)-5=-2.
  (2, 1) -- substitute x=2 into the equation and find y=3(2)-5=1.
  These are ordered pairs that satisfy the equation.

Examples:

Complete each ordered pair so that it satisfies the equation y = 8x - 1

   a. (0, y)   b. (-1, y)   c. (x, 0)   d. (x, -9)

Solutions: y = 8x - 1

   a. y = 8(0) - 1   b. y = 8(-1) - 1
      y = -1   y = -9
      (0 ,-1)   (-1, -9)

   c. 0 = 8(x) - 1   d. -9 = 8(x) - 1
      x = 1/8   x = -1
      (1/8, 0)   (-1, -9)

• To graph an ordered pair, consider the following:

y-axis

Q2 (Second quadrant) x⁻ y⁺	Q1 (First quadrant) x⁺ y⁺
Q3 (Third quadrant) x⁻ y⁻	Q4 (Fourth quadrant) x⁺ y⁻

x-axis

● The intersection of the x-axis and the y-axis is the origin (0, 0).

● The x-axis and the y-axis make up a rectangular coordinate system.

Example
Locate the points (0, 2), (-1, 3)
and (2, -3) on the graph.

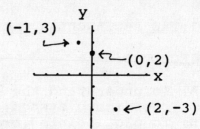

(B) To graph an equation, find the ordered pairs that satisfy the
equation and plot all these points on the graph.
● A straight line connects all these points.

Examples:
 Graph the equation y - 2x = -1.
 Plot at least 4 points.

        x-value: -1   0   1   2
        y-value: -3 -1   1   3

♦♦♦♦♦

EXERCISES

I. Complete the given ordered pairs for the equation.

        1. y = 2x - 5   (0,)   (, 5)
        2. 3y - x = 0   (-3,)   (, 6)
        3. -x = 2y + 4   (, -1)   (2,   )
        4. y = -3x + 90   (, 0)   (-30,)
        5. 5y = x   (-1,)   (, 1/5)
        6. x - y = 2   (, 0)   (5,   )
        7. 9x = 7y + 2   (, 1)   (0 ,)
        8. 2y = 3x - 7   (2,)   (, 1)

II. Plot the following points on a rectangular coordinate system and
    name the quadrant that it lies in or the axis that it lies on.

     9. (2, 3)   10. (-1, -2)   11. (2, -4)
    12. (0, -1)   13. (-3, 2)   14. (½, √3)
    15. (-2, -3)   16. (3.5, 0)   17. (0, 0)

III. Graph each of the following equations.
     Plot at least 4 points.

    18. x + y = 3   19. 3x + y = 0   20. 2x - 3y = 4

♦♦♦♦♦

178

**SOLUTIONS TO EXERCISES

I. Complete the given ordered pairs for the equation.

1. y = 2x - 5   (0, **-5**)   (5, 5)
2. 3y - x = 0   (-3, **-1**)   (18, 6)
3. -x = 2y + 4   (**-2**, -1)   (2, -3)
4. y = -3x + 90   (30, 0)   (-30, **180**)
5. 5y = x   (-1, **-1/5**)   (**1**, 1/5)
6. x - y = 2   (2, **0**)   (5, **3**)
7. 9x = 7y + 2   (**1**, 1)   (0, **-2/7**)
8. 2y = 3x - 7   (2, **-½**)   (**3**, 1)

II. Plot the following points on a rectangular coordinate system and
name the quadrant that it lies in or the axis that it lies on.

9. (2, 3)   **Q1**
10. (-1, -2)   **Q3**
11. (2, -4)   **Q4**

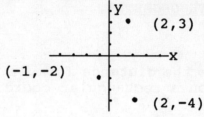

12. (0, -1)   **y-axis**
13. (-3, 2)   **Q2**
14. (½, √3)   **Q1**

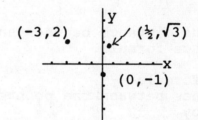

15. (-2, -3)   **Q3**
16. (3.5, 0)   **x-axis**
17. (0, 0)   **origin**

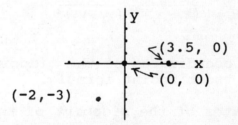

III. Graph each of the following equations. Plot at least 4 points.

18. x + y = 3

```
x :   -1   0   1   2
y :    4   3   2   1
```

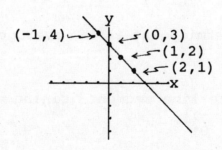

179

19. 3x + y = 0

x : 0   1   2   -1
y : 0 -3 -6   3

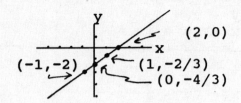

20. 2x - 3y = 4

x : -1   0   1   2
y : -2 -4/3 -2/3   0

♦♦♦♦♦

8.2 **COORDINATE GEOMETRY**

OBJECTIVES:

*(A) To define the distance formula between two points and the midpoint formula on a rectangular coordinate system.

** **REVIEW**

*(A) To find the distance between any two points in the system, use the distance formula.

Distance formula ---
The distance between the points $A(x_1, y_1)$ and $B(x_2, y_2)$ is given by

$$AB = \sqrt{(x_1 - x_2)^2 + (y_1 - y_2)^2}$$

● To find the coordinates of the midpoint of a line segment in the system, use the midpoint formula.

The coordinates of the midpoint of the line segment joining (x_1, y_2) and (x_2, y_2) are

$$\left(\frac{x_1 + x_2}{2}, \frac{y_1 + y_2}{2}\right)$$

●The x-coordinate of the midpoint is the average of the x-coordinates of the two endpoints.

●The y-coordinate of the midpoint is the average of the y-coordinates of the two endpoints.

Examples:
Find the midpoint of the line segment joining
    a. (-2, 4) and (5, -1)
    b. (0, 5) and (-6, 0)

c. (-10, 10) and (-10, 12)

Solutions:

For any two points (x_1, y_1) and (x_2, y_2)

midpoint formula $= (\dfrac{x_1+x_2}{2}, \dfrac{y_1+y_2}{2})$

a. (-2, 4) and (5, -1)   midpoint formula

$(x, y) = (\dfrac{-2+5}{2}, \dfrac{4+(-1)}{2}) = (\dfrac{-3}{2}, \dfrac{3}{2})$

b. (0, 5) and (-6, 0)

$(x, y) = (\dfrac{0+(-6)}{2}, \dfrac{5+0}{2}) = (-3, \dfrac{5}{2})$

c. (-10, 10) and (-10, 12)

$(x, y) = (\dfrac{-10+(-10)}{2}, \dfrac{10+(12)}{2}) = (-10, 11)$

● Use the distance formula and the midpoint formula of a line segment to solve geometric problems.

Examples:

a. Show that the points A(5, 3), B(5, -1), and C(1, -1) are vertices of an isosceles triangle.

Distance formula $= \sqrt{(x_2-x_1)^2+(y_2-y_1)^2}$

$AB = \sqrt{(5-5)^2+(3-(-1)^2} = 4$

$AC = \sqrt{(5-1)^2+(3-(-1))^2} = \sqrt{32}$

$BC = \sqrt{(5-1)^2+((-1)-(-1))^2} = 4$

$AB = BC$

Since the length of two sides of a triangle are equal, this triangle is isosceles.

b. Find the midpoint of the line segment joining (5, -6) and (-3, 4).

$(x, y) = (\dfrac{x_1+x_2}{2}, \dfrac{y_1+y_2}{2})$ midpoint formula

($x_1=5$, $x_2=-3$ and $y_1=-6$, $y_2=4$)

$(x, y) = (\dfrac{5+(-3)}{2}, \dfrac{(-6)+(4)}{2}) = (1, -1)$

◆◆◆◆◆

I. Find the distance between each pair of points.

   1. (5, 0), (-1, -2)   2. (-1, 3), (4, 0)   3. (5, 3), (-1, -1)
   4. (0.5, -5), (4, -1.5)   5. (0, 0), (5, -1)   6. $(\sqrt{2}, \sqrt{3}), (\sqrt{2}, 5\sqrt{3})$
   7. (-1, -1), (-2, -2)   8. (0, 5), (5, 0)   9. $(\frac{1}{2}, \frac{1}{4}), (-\frac{1}{4}, -\frac{1}{2})$

II. Find the midpoint of the line segment with the given endpoints.

   10. (5, 0), (-1, -2)   11. (-1, 3), (4, 0)   12. (0.5, -5),(-2,2)
   13. (5, 3), (-1, -1)   14. (0, 0), (5, -1)   15. (4, -1.5),(1, 2)

III. Solve each geometric problem.

   16. Determine whether or not the points (1, 2), (4, 0) and (-1, -2) are the vertices of an isosceles triangle.

   17. The lengths of the two sides of a right triangle are 3 and 4. What is the length of the hypotenuse?

   18. Determine whether or not the points (0, -10, (3, 0), and (-1,-2) lie on the same straight line.

   19. Determine whether or not the points (0, 0), (0, 5), (5, 0) and (5, 5) are the vertices of a square.

   20. A 10 foot pole is leaning against a wall. The bottom of the pole is 8 feet from the wall. How far will the top of the pole be above the ground?

♦♦♦♦♦

SOLUTIONS TO EXERCISES

I. Find the distance between each pair of points.

   1. (5, 0), (-1, -2) $d = 2\sqrt{10}$ 2. (-1, 3), (4, 0) $d = \sqrt{34}$

   3. (5, 3), (-1, -1) $d = 2\sqrt{13}$ 4. (0.5,-5),(4,-1.5) $d = \sqrt{24.5}$

   5. (0, 0), (5, -1) $d = \sqrt{26}$ 6. $(\sqrt{2}, \sqrt{3}), (\sqrt{2}, 5\sqrt{3})$ $d = 4\sqrt{3}$

   7. (-1, -1), (-2, -2) $d = \sqrt{2}$ 8. (0, 5), (5, 0) $d = 5\sqrt{2}$

   9. $(\frac{1}{2}, \frac{1}{4}), (-\frac{1}{4}, -\frac{1}{2})$ $d = \frac{3}{4}\sqrt{2}$

II. Find the midpoint of the line segment with the given endpoints.

$Midpoint\ formula = (x, y) = (\dfrac{x_1 + x_2}{2}, \dfrac{y_1 + y_2}{2})$

10. (5, 0), (-1, -2)
    $x_1 = 5$, $x_2 = -1$
    $y_1 = 0$, $y_2 = -2$
    midpoint = (2, -1)

11. (-1, 3), (4, 0)
    $x_1 = -1$, $x_2 = 4$
    $y_1 = 3$, $y_2 = 0$
    midpoint = (3/2, 3/2)

12. (0.5, -5), (-2, 2)
    $x_1 = 0.5$, $x_2 = -2$
    $y_1 = -5$, $y_2 = 2$
    midpoint = (-3/4, -3/2)

13. (5, 3), (-1, -1)
    $x_1 = 5$, $x_2 = -1$
    $y_1 = 3$, $y_2 = -1$
    midpoint = (2, 1)

14. (0, 0), (5, -1)
    $x_1 = 0$, $x_2 = 5$
    $y_1 = 0$, $y_2 = -1$
    midpoint = (5/2, -1/2)

15. (4, -1.5), (1, 2)
    $x_1 = 4$, $x_2 = 1$
    $y_1 = -1.5$, $y_2 = 2$
    midpoint = (5/2, 1/4)

III. Solve each geometric problem.

16. Determine whether or not the points A(1, 2), B(4, 0) and C(-1, -2) are the vertices of an isosceles triangle.

$AB = \sqrt{(1-4)^2 + (2-0)^2} = \sqrt{13}$ *distance between* (1, 2) *and* (4, 0)

$BC = \sqrt{(4-(-1))^2 + (0-(-2))^2} = \sqrt{29}$ *distance between* (4, 0) *and* (-1, -2)

$AC = \sqrt{(1-(-1))^2 + (2-(-2))^2} = \sqrt{20}$ *distance between* (1, 2) *and* (-1, -2)

This is not an isosceles triangle.

17. The lengths of the two sides of a right triangle are 3 and 4. What is the length of the hypotenuse?

$(AB)^2 = (3)^2 + (4)^2 = (9) + (16) = 25 = 5^2$
The hypotenuse is 5.

18. Determine whether or not the points A(0, -10), B(3, 0), and C(-1, -2) lie on the same straight line.
    (If these points lie on the same line, the sum of the length of the two shorter line segments should equal the length of the longest line segment)

$AB = \sqrt{(0-3)^2 + (-10-0)^2} = \sqrt{109}$ *distance between* (0, -10) *and* (3, 0)

$BC = \sqrt{(3-(-1))^2 + (0-(-2))^2} = \sqrt{20}$ *distance between* (3, 0) *and* (-1, -2)

$AC = \sqrt{(0-(-1))^2 + (-10-(-2))^2} = \sqrt{65}$ *distance between* (0, -10) *and* (-1, -2)

Since AB ≠ BC + AC. These points do not lie on the same line.

19. Determine whether or not the points (0, 0), (0, 5), (5, 0) and (5, 5) are the vertices of a square.

$AB = \sqrt{(0-0)^2 + (0-5)^2} = 5$ *distance between* (0,0) *and* (0, 5)

$AC = \sqrt{(0-5))^2 + (0-0)^2} = 5$ *distance between* (0,0) *and* (5,0)

$BD = \sqrt{(0-5)^2 + (5-5)^2} = 5$ *distance between* (0,5) *and* (5,5)

$CD = \sqrt{(5-5)^2 + (0-5)^2} = 5$ *distance between* (5,0) *and* (5,5)

     This is a square.

20. A 10 foot pole is leaning against a wall. The bottom of the pole is 8 feet from the wall. How far will the top of the pole be above the ground?
Apply the Pythagorean Theorem

$$(AB)^2 = (BC)^2 + (AC)^2$$
$$(10)^2 = 8^2 + (AC)^2$$
$$(AC)^2 = 10^2 - 8^2$$
$$(AC) = 100 - 64$$
$$AC = 6 \quad \text{The pole is 6 feet above the ground.}$$

♦♦♦♦♦

8.3 SLOPE

OBJECTIVES:

*(A) To understand the concept of the slope and how it relates to parallel lines and perpendicular lines.

*(B) To learn geometric applications.

** REVIEW

*(A) If (x_1, y_1) and (x_2, y_2) are any two distinct points, then the slope of the line joins through these points is defined as

$$m = \frac{y_2 - y_1}{x_2 - x_1} = \frac{change\ in\ y}{change\ in\ x}, \ where \ x_2 - x_1 \neq 0$$

Example:
Find the slope of a line containing the points (2, 3) and (-1, 5).

Solution:

$$m = \frac{y_2 - y_1}{x_2 - x_1} = \frac{(5-3)}{(-1)-(2)} = -\frac{2}{3}$$

where $x_1 = 2, x_2 = -1$ *and* $y_1 = 3, y_2 = 5$.

The same slope is obtained even if two points from the example are re-labelled.

$$m = \frac{y_2 - y_1}{x_2 - x_1} = \frac{3-5}{2-(-1)} = -\frac{2}{3}$$

where $x_1 = -1, x_2 = 2$ *and* $y_1 = 5, y_2 = 3$.

Therefore, the slope of the line through two points is also $-(2/3)$.

● **Parallel lines --**
Two lines are parallel if and only if their **slopes** are **equal**.

● **Perpendicular lines --**
Two lines are perpendicular if and only if **the product of their slopes is -1.**

Examples:

   a. Find the slope of the line passing through the point (2, -5) that is parallel to the line through (-1, 0) and (0, 3).

   b. Find the slope of the line passing through the point (2, -5) that is perpendicular to the line through (-1, 0) and (0, 3).

Solutions:
   a. The slope of the line through (-1, 0) and (0, 3) is 3. Therefore the slope parallel to the line through (-1, 0) and (0,3) should have the same slope 3.

   b. The slope of the line through (-1, 0) and (0, 3) is 3. If m is the slope perpendicular to the line through (-1, 0) and (0, 3), then m·3 = -1 implies m = -1/3.

● The slope of a vertical line is undefined and the slope of a horizontal line is 0.

◆◆◆◆◆

**EXERCISES

I. Find the slope of the line that goes through each of the following pairs of points.

   1. (0, -1), (2, 0)   2. (-1, 1), (3, 10)   3. (0, 5), (-4, ½)

   4. (4, -1), (5, -2)   5. (0, 0), (-1, -2)   6. (½, ¼), (3, -4)

   7. (8, -9), (9, -10)   8. (-1, -2), (3, 2)   9. (0.5, 0.3), (4, 1)

II. Find the distance and the midpoint between each of the following pairs of points.

10. (0, -1), (2, 0)   11. -1, 1), (3, 10)   12. (-1, -2), (3,2)

13. (0, 0), (-1, -2)   14. (0.5, 0.4), (4.5, 2.4)

15. (0, 5½), (-4, ½)

III. Determine if each of the following groups of ordered pairs
represents a right triangle.

16. (3, 5), (3, -2) , (8, -2)
17. (0, 0), (-1, -1), (2, -2)
18. (-1, 1), (0, -3), (-4, -2)
19. What is the slope of a line perpendicular to a line with a slope
of -½ ?
20. What is the slope of a line that goes through (-3, 1) and runs
parallel to the line through (-1, -5) and (0, 4)?

◆◆◆◆◆

**SOLUTIONS TO EXERCISES

I. Find the slope of the line that goes through each of the following
pair of points.

$$Slope = \frac{y_2 - y_1}{x_2 - x_1}$$

1. (0, -1), (2, 0)   m = ½   2. (-1, 1), (3, 10)   m = 9/4
3. (0, 5), (-4,½)   m = 9/8   4. (4, -1), (5, -2)   m = -1
5. (0, 0), (-1, -2)   m = 2   6. (½, ¼), (3, -4)   m = -17/10
7. (8, -9), (9, -10)   m = -1   8. (-1, -2), (3, 2)   m = 1
9. (0.5, 0.3),(4,1)   m = 1/5

II. Find the distance and the midpoint between each of the following
pair of points.

10. (0,-1), (2,0)   d = $\sqrt{5}$; midpoint = (1, -½).
11. (-1,1), (3,10)   d = $\sqrt{97}$; midpoint = (1, 11/2).
12. (-1,-2), (3,2)   d = 4$\sqrt{2}$; midpoint = (1, 0).
13. (0,0), (-1,-2)   d = $\sqrt{5}$; midpoint = (-½, -1).
14. (0.5,0.4), (4.5,2.4)   d = 2$\sqrt{5}$; midpoint = (2.5, 1.4).
15. (0,5½), (-4,½)   d = $\sqrt{41}$; midpoint = (-2, 3).

III. Determine if each of the following groups of ordered pairs
represents a right triangle.

Pythagorean Theorem:
If the sum of the squares of the legs of a triangle is equal to the
hypotenuse squared, then it is a right triangle.

186

16. A(3, 5), B(3, -2) , C(8, -2)

    AB = 7, BC = 5, and AC = $\sqrt{74}$
    $(AC)^2 = (AB)^2 + (BC)^2$
    This is a right triangle.

17. A(0, 0), B(-1, -1), C(2, -2)

    AB = $\sqrt{2}$, AC = $\sqrt{8}$, and BC = $\sqrt{10}$
    $(BC)^2 = (AB)^2 + (AC)^2$
    This is a right triangle.

18. A(-1, 1), B(0, -3), C(-4, -2)

    AB = $\sqrt{17}$, BC = $\sqrt{17}$, and AC = $\sqrt{18}$
    $(AC)^2 \neq (AB)^2 + (BC)^2$
    This is **not** a right triangle.

19. What is the slope of a line perpendicular to a line with a slope
    of $-\frac{1}{2}$?
    Two lines are perpendicular if and only if the product of their
    slopes is -1. Therefore $m_1 \cdot m_2 = -1$; $(-\frac{1}{2}) \cdot m_2 = -1$ implies $m_2 = 2$.

20. What is the slope of a line that goes through (-3, 1) and runs
    parallel to the line through (-1, -5) and (0, 4)?

    The slope of the line passes through (-1, -5) and (0, 4)
    is 9. The line parallel to it has the same slope m = 9.

♦♦♦♦♦

8.4 <u>EQUATIONS OF LINES</u>

<u>OBJECTIVES</u>:

*(A) To define the slope-intercept form and the standard form.
*(B) To graph the equation using the slope-intercept form.

** <u>REVIEW</u>

*(A) The equation of a line with slope m and y-intercept b is written in
    slope-intercept form --- y = mx + b.
    (0, b) is the point where the line crosses the y-axis.

Examples:
    a. Find the slope and the y-intercept of the line defined by the
       equation 6x - 3y = 15.

    b. Find the equation of a line with a slope of -2 and whose
       y-intercept is (0, 3).

Solutions:
    a. Express the equation 6x - 3y = 15 in slope-intercept form.
       y = 2x + (-5) --- divide both sides by 3.
       m = 2 and b = -5.

187

b. m = -2 and y-intercept is (0, 3) which implies b = 3
     y = mx + b = (-2)x + (3)
     y + 2x = 3

● Standard form --- ax + by = c, where a and b are not both 0.
  The equation of all lines can be written in the standard form.

Example:
     Write the equation of the line y = ½ x + 3 in standard form.

Solution:
     y = ½x + 3   -- subtract ½x from both sides
     y - ½x = 3   -- multiply both sides by 2
     2y - x = 6 --- re-arrange the equation so that it is in
                     standard form

(B) To graph a line using its slope and y-intercept.
    (y = mx + b)
    1. Locate the point (0, b).
    2. Locate the second point by using the rise and run technique.
    3. Draw a line through the two points.

Examples:
     a. Graph the equation 3x + 6y = 12.
     b. Write the equation in slope-intercept form for the line through
        (-4, 2) that is parallel to the line x + 4y = 1.

Solution:
  a. 3x - 6y = 12   rearrange the terms
     6y = 3x - 12   divide both sides by 6
     y = ½x - 2
     The line passes through (0, -2).
     The slope m = ½ = rise/run
     = go up (rise) 1 unit and to the right (run) 2 units.

     1. Locate the first point y-intercept (0, -2)
     2. Locate the second point by starting at (0,-2).
        Then go up 1 unit and to the right 2 units.
     3. Draw a line through the two points.

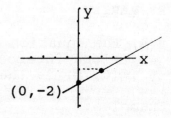

(0,-2)

  b. Find the slope of the equation x + 4y = 1 by re-arranging the
     equation in slope-intercept form.
     4y = -x + 1
     y = -¼x + ¼   where m_1 = -¼ and b = ¼
     m_1 = m_2 = -¼   two lines are parallel
     y = m_2x + b   substitute x = -4, y = 2, m_2 = -¼, and solve for b
     2 = -¼(-4) + b implies b = 2 - 1 = 1
     Therefore y = -¼x + 1.
                    ◆◆◆◆◆

EXERCISES

I. True or False?

1. The line $y = -x$ passes through the origin $(0, 0)$ and $(1, -1)$.

2. The line $y = x + 1$ is perpendicular to the line $y = x$.

3. The line $x = 2 + 3y$ has a slope of 3.

4. The equation $2x = 5$ has a graph that is a horizontal line and parallel to x-axis.

5. The equation $2x = 5$ has a graph that is a vertical line and parallel to y-axis.

6. The lines $3x + y = 1$ and $y + 3x = 1$ are parallel.

7. The lines $2x - 5y = 5$ and $2x - 5y = 7$ are perpendicular.

8. The lines $y = 5$ and $x = 10$ are perpendicular.

9. The equation $y = 8x + 10$ is in slope-intercept form and $(10, 0)$ is the y-intercept.

10. $x - 1 = 3(y - 2)$ is in slope-intercept form and the slope is -3.

II. Find the slope and the y-intercept for each of the following lines.

11. $2y = x + 1$ 12. $y = 3x - 5$ 13. $y = 8$

14. $x = 9 + y$ 15. $\frac{1}{4}x - \frac{1}{2}y = 1$ 16. $-3y = 2x + 6$

III. Write each equation in standard form and slope-intercept form.

17. $2y = x + 1$ 18. $y = 3x - 5$

19. $y = 8$ 20. $\frac{1}{4}x - \frac{1}{2}y = 1$

♦♦♦♦♦

SOLUTIONS TO EXERCISES

I. True or False ?

1. True.   2. False. ($m_1 = m_2 = 1$ and $m_1 \cdot m_2 \neq -1$)   3. False. ($m = 1/3$)

4. False. (This is a vertical line and parallel to y-axis)

5. True.   6. True. ($m_1 = -3$ and $m_2 = -3$)   7. False. ($m_1 \cdot m_2 \neq -1$)

8. True. ($y = 5$ is a horizontal line 5 units above the x-axis and $x = 10$ is a vertical line 10 units to the right of y-axis)

9. False. (The y-intercept is (0,10))

10. False. (Slope-intercept form is y = (1/3)x + 5/3 and m = 1/3)

II. Find the slope and the y-intercept for each of the following lines.

11. 2y = x + 1
    y = ½x + ½
    m = ½, (0, ½)

12. y = 3x - 5
    y = 3x + (-5)
    m = 3, (0,-5)

13. y = 8
    y = 0·x + 8
    m = 0, (0, 8)

14. x = 9 + y
    y = 1·x - 9
    m = 1, (0, -9)

15. ¼x - ½y = 1
    y = ½x - 2
    m = ½, (0, -2)

16. (-3y) = 2x + 6
    y = (-2/3)x -2
    m = -3/2, (0,-2)

III. Write each equation in standard form and slope-intercept form.

	standard form	slope-intercept form
17. 2y = x + 1	x - 2y = -1	y = ½x + ½
18. y = 3x - 5	3x - y = 5	y = 3x - 5
19. y = 8	y = 8	y = 0x + 8
20. ¼x - ½y = 1	x - 2y = 4	y = ½x - 2

◆◆◆◆◆

8.5 <u>MORE ON EQUATIONS OF LINES</u>

<u>OBJECTIVES</u>:

*(A) To define point-slope form.
*(B) To understand the slopes of perpendicular lines.
    To understand the slopes of parallel lines.

** <u>REVIEW</u>

*(A) Consider the slope of a line passing through (x_1, y_1).

The slope m of the equation is

$$m = \frac{y-y_1}{x-x_1}, \qquad where\ x \neq x_1$$

$(y-y_1) = m(x-x_2)$ *multiply both sides by* $(x-x_1)$.

This equation is called the point-slope form of the equation of a line.

Example:

Find the equation of the line that passes through (-2, 3) and whose slope is 3. Write it in slope-intercept form, standard form, and point-slope form.

Solution:

$$m = \frac{y - y_1}{x - x_1},$$ where $(x_1, y_1) = (-2, 3)$ and $m = 3$

$$3 = \frac{y - 3}{x - (-2)} = \frac{y - 3}{x + 2}$$ multiply both sides by $(x+2)$

$3(x+2) = y - 3$ $x \neq -2$

$y - 3 = 3(x+2)$ point-slope form $((y - y_1) = m(x - x_1))$

$3x - y = -9$ standard form $(ax + by = c)$

$y = 3x + 9$ slope-intercept form $(y = mx + b)$

*(B) Two lines are parallel if and only if their slopes are equal. Two lines are perpendicular if and only if the product of their slopes is -1.

Examples:
Write an equation of a line passing through the origin that is

a. perpendicular to 3x - 5y = 6,   b. parallel to 2x + 3y = 1.

Solutions:
a. 3x - 5y = 6 --- write the equation in slope-intercept form
   5y = 3x - 6
   y = (3/5)x - (6/5)   --- slope-intercept form

   If m_1 = 3/5, then m_2 = -(5/3) --- two lines are perpendicular
   (x - 0) = (-5/3)(y - 0)   --- the line passes through (0,0)
   3x = -5y   --- simplify
   y = -(3/5)x   --- slope-intercept form
   3x + 5y = 0   --- standard form

  b. 2x + 3y = 1
   3y = -2x + 1
   y = -(2/3)x + 1/3   --- slope-intercept form and m=(-2/3)
   (x - 0) = (-2/3)(y - 0) --- the line passes through (0,0) and has
                               a slope of (-2/3)
   3x = -2y
   3x + 2y = 0   --- standard form
   y = -(3/2)x   --- slope-intercept form

♦♦♦♦♦

**EXERCISES

I. Find which pairs of lines are parallel.
   Find which pairs of lines are perpendicular.

    1. 2x = y and 6x - 3y = 1
    2. 8x + y = 5 and x - 8y + 8 = 0
    3. The line through (3, 4) and (9, 6) and the line through
       (0, 1) and (-3, 0)
    4. 2x = y and 2y = x

5. $3x - 4y = 6$ and $4x + 3y = 1$
6. $2x + 3y = 5$ and $14x + 21y = 35$
7. $2x + y = 2$ and $y - x = 3$
8. $2y - x = 2$ and $y + 2x = 3$
9. $5y - x = 12$ and $y - 1/5x = 10$
10. $10x - 9y = 1$ and $x - y = 2$

II. Find the equation of each of the lines described below, and write the answer in point-slope and slope-intercept form.

11. The line containing the point $(2, -1)$ and parallel to the line $y = 2x + 3$

12. The line parallel to the line $2y -x = 9$ passing through the point $(-3, -3)$

13. The line perpendicular to the line $2y -x = 9$ passing through the point $(-3, -3)$

14. The line through the point $(-2, 5)$ with a slope of $\frac{1}{2}$

15. The line through the point $(1, 4)$ with a slope of -5

III. Write each equation in standard form.

16. $2y - 5 = -2(x - 5)$

17. $y - \frac{1}{2} = - 1/5(x - \frac{1}{2})$

Write each equation in slope-intercept form.

18. $2y - 5 = -2(x - 5)$

19. $(y - \frac{1}{2}) = -1/5(x - \frac{1}{2})$

20. $\frac{1}{4}(y - 2) = - \frac{1}{4}(x + 5)$

♦♦♦♦♦

SOLUTIONS TO EXERCISES

I. Find which pairs of lines are parallel.
Find which pairs of lines are perpendicular.

	parallel	perpendicular
1. $2x = y$ and $6x - 3y = 1$ $m_1 = 2$, $m_2 = 2$	yes	no
2. $8x + y = 5$ and $x - 8y + 8 = 0$ $m_1 = -8$, $m_2 = 1/8$	no	yes

192

3. The line through (3, 4) and (9, 6)   yes   no
   and the line through (0, 1) and (-3, 0)
   $m_1 = 1/3$, $m_2 = 1/3$

4. $2x = y$ and $2y = x$ no   no
   $m_1 = 2$, $m_2 = \frac{1}{2}$

5. $3x - 4y = 6$ and $4x + 3y = 1$ no   yes
   $m_1 = 3/4$, $m_2 = -4/3$

6. $2x + 3y = 5$ and $14x + 21y = 35$ yes   no
   $m_1 = -2/3$, $m_2 = -2/3$

7. $2x + y = 2$ and $y - x = 3$ no   no
   $m_1 = -2$, $m_2 = 1$

8. $2y - x = 2$ and $y + 2x = 3$ no   yes
   $m_1 = \frac{1}{2}$, $m_2 = -2$

9. $5y - x = 12$ and $y - 1/5x = 10$ yes   no
   $m_1 = 1/5$, $m_2 = 1/5$

10. $10x - 9y = 1$ and $x - y = 2$ no   no
    $m_1 = 10/9$, $m_2 = 1$

II. Find the equation of each of the lines described below, and write the answer in point-slope and slope-intercept form.

$y - y_1 = m(x - x_1)$ -- point-slope form, where m is the slope and (x_1, y_1) is the known point.

$y = mx + b$ -- slope-intercept form, where m is the slope and b is the y-intercept.

11. The line containing the point (2, -1) and parallel to the line $y = 2x + 3$.
    $(y-(-1)) = m(x-2)$ --- m = 2 since it is parallel to $y = 2x + 3$
    $y + 1 = 2(x - 2)$ --- point-slope form
    $y = 2x + (-5)$ --- slope-intercept form

12. The line parallel to the line $2y -x = 9$ passing through the point (-3, -3).
    $(y -(-3)) = m(x -(-3))$ --- $m = \frac{1}{2}$
    $y + 3 = \frac{1}{2}(x + 3)$ --- point-slope form
    $y = \frac{1}{2}x - 3/2$ --- slope-intercept form

13. The line perpendicular to the line $2y - x = 9$ passing through the point (-3, -3).
    The slope of the line $2y-x=9$ is $\frac{1}{2}$.
    $(y -(-3)) = m(x -(-3))$ --- $m_1 \cdot m_2 = -1$ and $m_1 = \frac{1}{2}$ implies $m_2 = -2$
    $y + 3 = (-2)(x + 3)$ --- point-slope form
    $y = -2x - 9$ --- slope-intercept form

14. The line through the point (-2, 5) with a slope of ½
    y - 5 = m(x -(-2))
    y - 5 = ½(x + 2)   --- point-slope form
    y = ½x + 6   --- slope-intercept form

15. The line through the point (1, 4) with a slope of -5
    y - 4 = m(x - 1)   --- m = -5
    y - 4 = -5(x - 1)   --- point-slope form
    y = -5x + 9   --- slope-intercept form

III. Write each equation in standard form.
     standard form --- ax + by = c

   16. 2y - 5 = -2(x - 5)   standard form   2x + 2y = 15
   17. y - ½ = -(1/5)(x - ½)   standard form   2x + 10y = 6

   Write each equation in slope-intercept form.
   slope-intercept form --- y = mx + b

   18. 2y - 5 = -2(x - 5)
       2y - 5 = -2x + 10
       y = (-1)x + 15/2, m = -1.

   19. (y - ½) = -1/5(x - ½)
       10[(y - ½)] = 10[-1/5(x - ½)]
       10y - 5 = -2x + 1
       10y = -2x + 6
       y = -(1/5)x + 3/5, m = -1/5.

   20. ¼(y - 2) = - ¼(x + 5)
        y - 2 = -(x + 5)
        y = -x -5 + 2
        y = (-1)x - 3, m = -1.

♦♦♦♦♦

8.6 **APPLICATIONS OF LINEAR EQUATIONS**

OBJECTIVES:

*(A) To find the linear relationship between two variables and to
     express it as a linear equation.

** **REVIEW**

*(A) Step 1. Read the problem carefully.
      Step 2. Understand all the given information.
      Step 3. Express the unknown(variable) in terms of all the given
              information.
      Step 4. Solve the variable from the equation.

♦♦♦♦♦

**EXERCISES

I. True or False?

1. If $d = t^2 + 5$, then there is a linear relationship between t and d.

2. There is a linear relationship between the area of a right triangle and its height.

3. There is a linear equation which expresses the volume of a cube in terms of the length of its side.

4. If $y = 3x + z^2 + w^2$, then there is a linear relationship between x and y.

5. If $y = 3x + z^2 + w^2$, then there is a linear relationship between x and w.

II. In each case write a linear equation that expresses one variable in terms of the other.

6. Express weight in pounds in terms of weight in ounces.

7. Express weight in ounces in terms of weight in pounds.

8. For a triangle with a fixed height of 10 feet, express the area in terms of its base.

9. Express the perimeter of a square in terms of its side s.

10. Express the Celsius temperature C in terms of the Fahrenheit temperature F.

♦♦♦♦♦

**SOLUTIONS TO EXERCISES

I. True or False ?

1. False.   2. True.   (Area = ½b·h)   3. False. (Volume = s^3)
4. True.   5. False.

II. In each case write a linear equation that expresses one variable in terms of the other.

6. p = 16 oz.   7. oz = (1/16)p   8. Area = ½·10(b)= 5b
9. p = 4s   10. C = (5/9)(F - 32)

♦♦♦♦♦

8.7 **FUNCTIONS**

OBJECTIVES:

*(A) To understand functions determined by a formula,
     and functions determined by a table.
*(B) To understand the domain and the range of a function.

** **REVIEW**

*(A) A function determined by a formula.
 Any formula where the value of one variable determines a unique value
 for another variable is called a function.

Examples:

    a. Express the area of a circle as a function of its radius.
 $C = \pi r^2$ --- every value of r there is a unique value of C.

    b. The formula for a circle $x^2 + y^2 = r^2$ is **not** a function.
 Because for every value of x there are two values of y.
 If x = 0, then y = ± r.

● A function determined by a table.

 If one value of a variable corresponds to only one value of another
 variable in the table, the first variable is a function of the second
 variable.

Examples:

a. x	y
3	0
4	1
5	0
6	1
2	12

b. p	q	
-1	8	<---
-1	9	<---
0	10	
1	11	

    a. x is a function of y because every value of x
 corresponds to only one value of y.

    b. p is **not** a function of q because
 p = -1 corresponds q = 8 and 9.

● "y is a function of x " means there is a relationship between x and y
 such that for every x there is one and only one value of y.

● Alternative definition of a function.
 A function is a set of ordered pairs of real numbers such that no two
 ordered pairs have the same first coordinates but two different
 second coordinates.

Examples:

    a. $y = x^2 + 3x + 1$
       y is a function of x because for every value of x there
       is one and only one y value.

    b. (3, 2), (5, 2), (0, 1), (5, 1) is **not** a function.
       When x = 5, y = 2 and y = 1.

*(B)
  The set of all the values of x is called the **domain** of the function.
  The set of all the values of y is called the **range** of the function.
  x is the independent variable and y is the dependent variable.

Examples:

    State the domain and the range of each function.

    a. $y = 2x^2 + 5$ b. $y = \sqrt{x} + 1$ c. {(3, 1) (4, 2) (5, 3)}

Solutions:

        a. $y = 2x^2 + 5$
          D = {x: x is a real number}
          R = {y: y ≥ 5}.

        b. $y = \sqrt{x} + 1$
          D = {x : x ≥ 0}
          R = {y : y ≥ 1}.

        c. {(3, 1) (4, 2) (5, 3)}
          D = {3, 4, 5}
          R = {1, 2, 3}.

    The symbol f(x) is often used as the notation to represent
    y as a function of x.

    y = f(x), where x is the independent variable and
    y is the dependent variable.

Example:
      $y = 3x^2 + 5x + 1$ can be written as $f(x) = 3x^2 + 5x + 1$.

Examples:

    a. If x = 0, then $f(0) = 3(0)^2 + 5(0) + 1 = 1$

      Therefore f(0) = 1.

b. $f(x) = x^3 - 1$ find $f(0)$, $f(-1)$, and $f(2)$

$$f(0) = (0)^3 - 1 = -1$$
$$f(-1) = (-1)^3 - 1 = -2$$
$$f(2) = (2)^3 - 1 = 7.$$

◆◆◆◆◆

EXERCISES

I. Find which sets of ordered pairs are functions.

1. {(1, 2), (2, 3), (3, 4)}
2. {(0, 0), (1, 1), (2, 2), (3, 3)}
3. {(2, -1), (2, -2), (3, -1), (4, -1)}
4. {(4, 4), (5, 5), (6, 6)}
5. {(1, 2), (1, 3), (1, 4)}
6. {(6, 6)}
7. { } empty set (The set contains no elements).

II. Determine whether or not each equation defines a function.

8. $2x = 3y$ 9. $y^2 = x^2 + 1$ 10. $y = \sqrt[5]{x}$
11. $y^2 = x$ 12. $x^2 = y$ 13. $7 = |x + 2|$

III. Determine the domain and the range of each function.

14. {(-2, 1) (3, 1) (5, 1)}   15. {(2, 0) (0, 2)}

16. $f(x) = x + 5$ 17. $f(x) = x^3 + 1$

18. $f(x) = |x + 1|$ 19. $f(x) = -3t^2$

20. $f(x) = x^2 + x + 1$

◆◆◆◆◆

SOLUTIONS TO EXERCISES

I. Find which sets of ordered pairs are functions.

1. {(1, 2), (2, 3), (3, 4)} is a function.
2. {(0, 0), (1, 1), (2, 2), (3, 3)} is a function.
3. {(2, -1), (2, -2), (3, -1), (4, -1)} is **not** a function.
4. {(4, 4), (5, 5), (6, 6)} is a function.
5. {(1, 2), (1, 3), (1, 4)} is **not** a function.
6. {(6, 6)} is a function.
7. { } = ∅ is a function.

II. Determine whether or not each equation defines a function.
   (y is a function of x)

   8. $2x = 3y$ is a function.

   9. $y^2 = x^2 + 1$ is **not** a function   (when $x = 0$, $y = \pm 1$).

   10. $y = {}^5\sqrt{x}$ is a function.

   11. $y^2 = x$ is **not** a function   (when $x = 4$, $y = \pm 2$).

   12. $x^2 = y$ is a function.

   13. $7 = |x + 2|$ is a function.

III. Determine the domain and the range of each function.

   14. $\{(-2, 1)\ (3, 1)\ (5, 1)\}$
       D = $\{-2, 3, 5\}$
       R = $\{1\}$

   15. $\{(2, 0)\ (0, 2)\}$
       D = $\{2, 0\}$
       R = $\{0, 2\}$

   16. $f(x) = x + 5$
       D = {x : x is a real number}
       R = {x : x is a real number}

   17. $f(x) = x^3 + 1$
       D = {x : x is a real number}
       R = {x : x is a real number}

   18. $f(x) = |x + 1|$
       D = {x : x is a real number}
       R = {y : y ≥ 0}

   19. $f(x) = -3t^2$
       D = {x : x is a real number}
       R = {y : y ≤ 0}

   20. $f(x) = x^2 + x + 1$
       D = {x : x is a real number}
       R = {x : x is a real number}

◆◆◆◆◆

8.8 **VARIATION**

OBJECTIVES:

*(A) To define direct variation, inverse variation and joint variation.

** <u>REVIEW</u>

*(A)
- If x varies **directly** as y, then x = k·y.
- If x varies **inversely** as y, then x = k·(1/y).
- If x varies **jointly** as y and z, then x = k·y·z.
  (k is a nonzero real number called a constant.)

Examples:

    a. If x varies directly as y^3, then $x = k \cdot y^3$.
    b. If x varies inversely as y^2, then $x = k/y^2$.
    c. If x varies jointly as y^2 and z^3, then $x = k \cdot (y^2 z^3)$.
    d. If x varies directly as y^4 and inversely as z^3,
       then $x = k \cdot (y^4/z^3)$.
    e. If x varies jointly as y^4 and inversely as z^3, then $x = k(y^4/z^3)$.

♦♦♦♦♦

**<u>EXERCISES</u>

I. Write a formula that expresses the relationship described by each statement.

    1. C varies directly as D.
    2. x varies directly as the square root of y.
    3. t varies directly as the fourth power of m.
    4. f varies directly as the product of t_1 and t_2.
       but inversely as cube root of r.
    5. p varies jointly as q and r.
    6. t varies directly as x^2.
    7. m varies inversely as x^3.
    8. a varies jointly as b and $\sqrt{c}$.
    9. r is directly proportional to $\sqrt{x}$.
    10. x varies directly as y and inversly as z.

II. Find the variation constant and write a formula that express the indicated variation.

    11. C varies directly as D, and C = 1 when D = 3.
    12. x varies directly as the square root of y,
       and x = 6 when y = 9.
    13. t varies directly as the fourth power of m, and t = 8
       when m = 2.
    14. f varies directly as the product of t_1 and t_2 but
       inversely as cube root of r , f = 12
       when $t_1 = 2$, $t_2 = 3$, and r = 1.
    15. p varies jointly as q and r, p = ¼ when q = -1 and r = -2.

III. Solve each variation problem.

    16. C varies directly as D, and C = 1 when D = 3. Find C when D = 0
    17. x varies directly as the square root of y, and x = 6 when y = 9.
       Find x when y = 2.

18. t varies directly as the fourth power of m, and t = 8 when m = 2. Find t when m = -1.

19. f varies directly as the product of t_1 and t_2 but inversely as the cube root of r. f = 12 when t_1 = 2, t_2 = 3, and r = 1. Find f when t_1 = 1, t_2 = 2 and r = -1.

20. p varies jointly as q and r, p = ¼ when q = -1 and r = -2. Find p when q = -1 and r = -2.

◆◆◆◆◆

SOLUTIONS TO EXERCISES

I. Write a formula that expresses the relationship described by each statement.

1. C varies directly as D. $C = kD$

2. x varies directly as the square root of y. $x = k\sqrt{y}$

3. t varies directly as the fourth power of m. $t = km^4$

4. f varies directly as the product of t_1 and t_2 but inversely as the cube root of r. $f = (kt_1t_2)/\sqrt[3]{r}$

5. p varies jointly as q and r. $p = kqr$

6. t varies directly as x^2. $t = kx^2$

7. m varies inversely as x^3. $m = k/x^3$

8. a varies jointly as b and $\sqrt{c}$. $a = k \cdot b \cdot \sqrt{c}$

9. r is directly proportional to $\sqrt{x}$. $r = k\sqrt{x}$

10. x varies directly as y and inversely as z. $x = ky/z$

II. Find the variation constant and write a formula that expresses the indicated variation.

11. C varies directly as D, and C = 1 when D = 3.
 C = (1/3)D (since 1 = k· (3) implies k = 1/3)

12. x varies directly as the square root of y, and x = 6 when y = 9.
 $x = 2\sqrt{y}$ (since 6 = k$\sqrt{9}$ implies k = 2)

13. t varies directly as the fourth power of m, and t = 8 when m = 2.
 $t = ½m^4$ (since 8 = k(2)^4 implies k = ½)

14. f varies directly as the product of t_1 and t_2 but inversely as the cube root of r, f = 12 when t_1 = 2, t_2 = 3, and r = 1.
 $f = (2t_1t_2)/\sqrt[3]{r}$ (since (12) = k(2)(3)/$\sqrt[3]{1}$ implies k = 2)

201

15. p varies jointly as q and r, p = ¼ when q = -1 and r = -2.
    p = (1/8)qr   (since (¼) = k(-1)(-2) implies k = 1/8)

III. Solve each variation problem.

16. C varies directly as D, and C = 1 when D = 3. Find C when D = 0.
    C = 0   (since k=1/3 and C = (1/3)D)

17. x varies directly as the square root of y, and x = 6 when y = 9.
    Find x when y = 2.
    x = 2$\sqrt{2}$ (since k=2 and x = 2$\sqrt{y}$)

18. t varies directly as the fourth power of m, and t = 8 when
    m = 2. Find t when m = -1.
    t = ½   (since k=½ and t = ½m^4)

19. f varies directly as the product of t_1 and t_2 but inversely as
    the cube root of r. f = 12 when t_1 = 2, t_2 = 3, and r = 1.
    Find f when t_1 = 1, t_2 = 2 and r = -1.
    f = -4   (since k=2 and f= 2($t_1 \cdot t_2$)/$\sqrt[3]{r}$)

20. p varies jointly as q and r, p = ¼ when q = -1 and r = -2.
    Find p when q = -1 and r = -2.
    p = ¼   (since k = 1/8 and p = (1/8)qr)

♦♦♦♦♦

CHAPTER VIII - TEST A

I. For each point, name the quadrant that it lies in or the axis that it
   lies on.

   1. (-1,-2)   2. (1, 2)   3. (-1, 3)
   4. (3, -1)   5. (0, 8)   6. (-2, 0)
   7. (0, 0)

II. Complete the ordered pairs so that each ordered pair satisfies
    the given equation.

    8. x + 5y = 6 ;   (0,   ),   (, 1)

    9. 2x - 7y = 9;   (0,   ),   (, -1)

    10. x + y + 2 ;   (1,   ),   (, 1)

    11. 2x + 3y = 4 ;   (-1,),   (, 2)

    12. 9x - 7y = 1 ;   (1/9,),   (, 0)

    13. 3x + 5y = 2 ;   (5,   ),   (, 3)

III. Use the distance formula to determine whether or not the following
     points lie on the same line.

14. (0, 1), (1, 2), (2, 3)   15. (0, 6), (4, 9), (-8, 0)

16. (1, 3), (-1, 2), (9, 7)   17. (0, 0), (2, -3), (4, 5)

18. (0, 1), (-1, 2), (3, -2)   19. (0, 0), (2, 2), (3, 3)

20. (2, 0), (-1, -2), (3, 0)

◆◆◆◆◆

CHAPTER VIII - TEST B

I. Write an equation in slope-intercept form for each of the line below.

1.
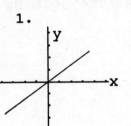

2.
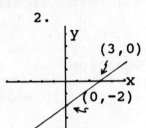

3.
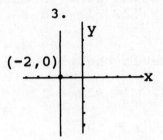

4.
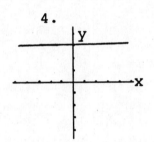

5.
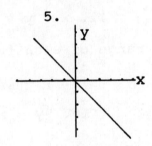

6.
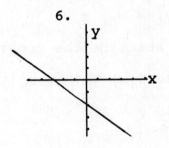

7.
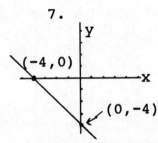

8.
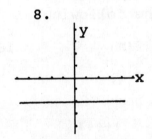

9.
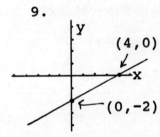

II. For each point, name the quadrant it lies in or the axis it lies on.

10. (-1, -2)   11. ($\sqrt{4}$, 3)=(2,3)   12. (3, -4)

14. (-½, 2)   15. (0, -8)   16. (π, -5)

III. Find the midpoint and the slope of the line segment
    joining each pair of points.

17. (-1, 0), (0, -4)   18. (3, 1), (-5, 6)

19. (-1, -2), (-3, -4)   20. (0, 0), (5, -5)

◆◆◆◆◆

CHAPTER VIII - TEST C

I. Write the equation of each line described below.

1. The line through (0, 0) with a slope of -2

2. The line through (-1, 2) with a slope of ½

3. The line through (-1, 2) parallel to 2x + y = 1

4. The line through (-3, 0) perpendicular to 2x + y = 1

5. The line through (2, -1) perpendicular to the line joining (2, -1) and (5, 1).

II. Sketch the graph of each equation.

6. y = -2x + 3

7. 2y = 3x - 4

8. y = -5

9. x = -5

10. x + y = 0

11. x - y = 0

III. Determine the domain and range of each function.

12. $f(x) = |x + 1| + 2$

13. y = 1/x

14. $y = 3\sqrt{x} + 5$

15. 9y + x = 0

IV. Let f(x) = 3x + 5. Find the following:

16. f(-1)

17. f(π)

18. f(2)

19. f(10)

20. f(½)

♦♦♦♦♦

CHAPTER VIII - TEST A - SOLUTIONS

I. For each point, name the quadrant that it lies in or the axis that it lies on.

1. (-1,-2)   **Q3**

2. (1, 2)   **Q1**

3. (-1, 3)   **Q2**

4. (3, -1)   **Q4**

5. (0, 8)   **on the y-axis**

6. (-2, 0) **on the x-axis**

7. (0, 0)   **origin**

II. Complete the ordered pairs so that each ordered pair satisfies the given equation.

8. $x + 5y = 6$;   (0, **6/5**),   (1, 1)
9. $2x - 7y = 9$;   (0, **-9/7**), (1, -1)
10. $x + y = 2$;   (1, **1**),   (3, -1)
11. $2x + 3y = 4$;   (-1, **2**),   (**-1**, 2)
12. $9x - 7y = 1$;   (1/9, **0**), (-13/9, -2)
13. $3x + 5y = 2$;   (-1, **1**),   (4, -2)

III. Use the distance formula to determine whether or not the following
points lie on the same line.

14. A(0, 1), B(1, 2), C(2, 3)

AB = $\sqrt{2}$, BC = $\sqrt{2}$, AC = $2\sqrt{2}$
AC = AB + BC
They lie on the same line.

15. A(0, 6), B(4, 9), C(-8, 0)

AB = 5, BC = 15, AC = 10
BC = AC + AB
They lie on the same line.

16. A(1, 3), B(-1, 2), C(9, 7)

AB = $\sqrt{5}$, AC = $4\sqrt{5}$, BC = $5\sqrt{5}$
BC = AB + AC
They lie on the same line.

17. A(0, 0), B(2, -3), C(4, 5)

AB = $\sqrt{13}$, BC = $2\sqrt{17}$, AC = $\sqrt{41}$
BC $\neq$ AB + AC
They are not on the same line.

18. A(0, 1), B(-1, 2), C(3, -2)

AB = $\sqrt{2}$, AC = $3\sqrt{2}$, BC = $4\sqrt{2}$
BC = AC + AB
They lie on the same line.

19. A(0, 0), B(2, 2), C(3, 3)

AB = $2\sqrt{2}$, BC = $\sqrt{2}$, AC = $3\sqrt{2}$
AC = BC + BC
They lie on the same line.

20. A(2, 0), B(-1, -2), C(3, 0)

AB = $\sqrt{13}$, BC = $2\sqrt{5}$, AC = 1
BC $\neq$ AB + AC
They are not on the same line.

♦♦♦♦♦

CHAPTER VIII - TEST B - SOLUTIONS

I. Write an equation in slope-intercept form for each of the lines
below.

1.

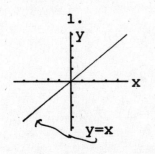

2.

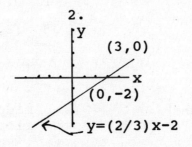

3.

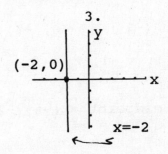

205

4.

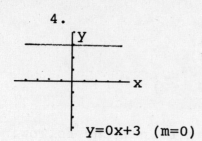

y=0x+3 (m=0)

5.

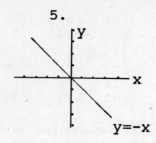

y=-x

6.

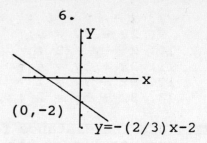

(0,-2)

y=-(2/3)x-2

7.

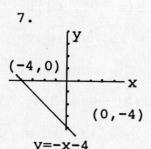

(-4,0)

(0,-4)

y=-x-4

8.

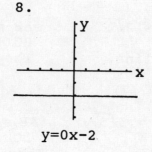

y=0x-2

9.

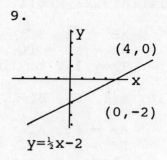

(4,0)

(0,-2)

y=½x-2

II. Locate the following points on a rectangular coordinate system.

10. (-1, -2)
11. ($\sqrt{4}$, 3)=(2,3)
12. (3, -4)

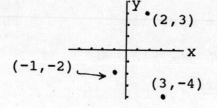

(-1,-2) →

(2,3)

(3,-4)

14. (-½, 2)
15. (0, -8)
16. (π, -5)

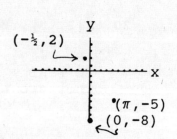

(-½,2)

(π,-5)

(0,-8)

III. Find the midpoint and the slope of the line segment that joins each pair of points.

midpoint formula = $(\frac{x_1+x_2}{2}, \frac{y_1+y_2}{2})$ and m = $\frac{y_2-y_1}{x_2-x_1}$

17. (-1, 0), (0, -4)

$x_1 = -1$, $x_2 = 0$
$y_1 = 0$, $y_2 = -4$

midpoint = (-½, -2)
slope = -4

18. (3, 1), (-5, 6)

$x_1 = 3$, $x_2 = -5$
$y_1 = 1$, $y_2 = 6$

midpoint = (-1, 7/2)
slope = -5/8

19. (-1, -2), (-3, -4)

$x_1 = -1$, $x_2 = -3$
$y_1 = -2$, $y_2 = -4$

midpoint = (-2, -3)
slope = 1

20. (0, 0), (5, -5)

$x_1 = 0$, $x_2 = 5$
$y_1 = 0$, $y_2 = -5$

midpoint = (5/2, -5/2)
slope = -1

♦♦♦♦♦

CHAPTER VIII - TEST C - SOLUTIONS

I. Write the equation of each line described below.

1. The line through (0, 0) with a slope of -2
   $(y - 0) = (-2)(x - 0)$
   $y = -2x$
   $y + 2x = 0$

2. The line through (-1, 2) with a slope of ½
   $(y - 2) = (½)(x - (-1))$
   $y - 2 = ½x + ½$
   $2y - x = 5$

3. The line through (-1, 2) and parallel to $2x + y = 1$
   $(y - 2) = (-2)(x -(-1))$
   $y - 2 = -2x - 2$
   $y + 2x = 0$

4. The line through (-3, 0) and perpendicular to $2x + y = 1$
   $(y - 0) = (½)(x -(-3))$
   $y = ½x + 3/2$
   $2y - x = 3$

5. The line through (2, -1) perpendicular to the line joining (2, -1) and (5, 1).

   The slope of the line joining (2, -1) and (5, 1) is 2/3
   Therefore the slope of the line perpendicular to it is -3/2
   $(y - (-1)) = (-3/2)(x - 2)$
   $y + 1 = (-3/2)x + 3$
   $2y + 3x = 4$

II. Sketch the graph of each equation

6. $y = -2x + 3$

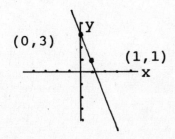

(0,3)

(1,1)

7. $2y = 3x - 4$
   $y = (3/2)x - 2$

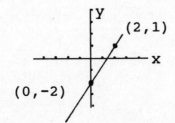

(2,1)

(0,-2)

8. $y = -5$

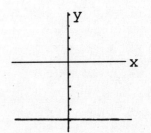

207

9. x = -5

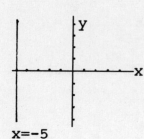

x=-5

10. x + y = 0
    y = -x

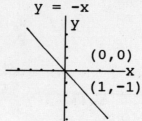

(0,0)
(1,-1)

11. x - y = 0
    y = x

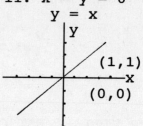

(1,1)
(0,0)

III. Determine the domain and range of each function.

12. f(x) = |x + 1| + 2
    D = { x: x is a real}
    R = { y: y ≥ 2}

13. y = 1/x
    D = {x: x ≠ 0}
    R = {y: y ≠ 0}

14. y = 3√x + 5
    D = {x: x ≥ 0}
    R = {y: y ≥ 5}

15. 9y + x = 0
    D = {x: x is a real}
    R = {y: y is a real}

IV. Let f(x) = 3x + 5. Find the following:

16. f(-1) = 3(-1) + 5 = 2
17. f(π) = 3(π) + 5
18. f(2) = 3(2) + 5 = 11
19. f(10) = 3(10) + 5 = 35
20. f(½) = 3(½) + 5 = 6.5

♦♦♦♦♦

CHAPTER 9

SYSTEMS OF EQUATIONS AND INEQUALITIES IN TWO VARIABLES

9.1 SOLVING SYSTEMS OF LINEAR EQUATIONS BY GRAPHING

OBJECTIVES:

*(A) To solve a system of equations by graphing.
*(B) To study three different types of systems:
    independent, inconsistent, and dependent.

** **REVIEW**

*(A) A pair of equations is called a system of equations.
    $a_1x + b_1y = c_1$
    $a_2x + b_2y = c_2$ where a and b are both not zero.

Examples:
    a. $2x - 3y = 5$ b. $\frac{1}{2}x - \frac{1}{4}y = -\frac{1}{4}$
       $x + y = 0$ $0.5x + 0.8y = -1.4$

These are examples of systems of equations.

● A pair of values for x and y which satisfies both equations is called
  a solution to the system.

Example:

    $2x - 3y = 5$ --> (1) refer to this equation as eq.(1)
    $x + y = 0$ --> (2) refer to this equation as eq.(2)

    If x=1 and y=-1, $2x-3y=2(1)-3(-1)=5$ -- correct (from eq.(1))
    If x=1 and y =-1, $x+y=1+(-1)=0$ -- correct (from eq.(2))
    Therefore (1, -1) is a solution to the system.

    If x=4 and y=1, $2x-3y=2(4)-3(1)=5$ -- correct (from eq.(1))
    If x=4 and y=1, $x+y=4+(1)=5 \neq 0$ -- incorrect (from eq.(2))
    Therefore (4, 1) is not a solution to the system.

● A point must satisfy **both equations** to be a solution of the system.

● Solve the system by graphing --

  Rewrite the equations in slope-intercept form for easy graphing.
  Graph both equations, obtaining two straight lines.
  The solution is the intersection of these two lines.

209

Examples: Solve the system by graphing.
  a. 2x -3y = 5   b. ½x - ¼y = -¼
     x + y = 0   0.5x + 0.8y = -1.4

Solutions:

      a. 2x -3y = 5 -- (1)
         x + y = 0 -- (2)

Write the system in slope-intercept form for easy graphing.

y = 2/3x - 5/3 --> from eq.(1), m = 2/3 and y-intercept is (0, -5/3)
y = -x + 0   --> from eq.(2), m = -1 and y-intercept is (0, 0)

The intersection of the
two lines is (1, -1).

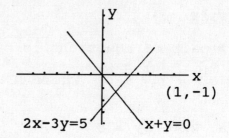

2x-3y=5   x+y=0

      b. ½x - ¼y = -¼   -- (1)
         0.5x + 0.8y = -1.4   -- (2)

         2x - y = -1   --   multiply both sides by 4 (from eq.(1))
         5x + 8y = -14 --   multiply both sides by 10 (from eq.(2))

      Express the above equations in slope-intercept form
      y = 2x + 1   -- m = 2 and passes through (0, 1)
      y = -5/8x - 7/4 -- m = -5/8 and passes through (0,-7/4)

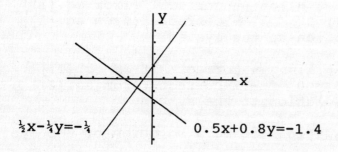

½x-¼y=-¼   0.5x+0.8y=-1.4

The intersection of the two lines is (-22/21, -23/21).

*(B)

- Inconsistent equations --
  If two lines are parallel, there is no point in common.
  Thus there is no solution.

- Dependent equations --
  If two equations are represented by the same line, then the two lines
  coincide, which means there is no unique solution.

- Independent equations --
  If two lines intersect at exactly one point, then there is exactly one
  solution to the system.

Examples:

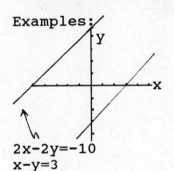

2x-2y=-10
x-y=3

(Inconsistent eq.)
 no solution

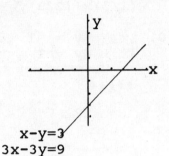

x-y=3
3x-3y=9

(Dependent eq.)
 no solution

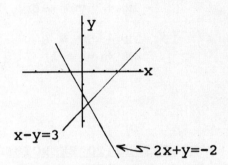

x-y=3
2x+y=-2

(Independent eq.)
 one unique solution
 The solution is (1/3, -8/3).

♦♦♦♦♦

**EXERCISES

I. Determine which of the given points is a solution to each system of
   equations.

1. (-1, 2), (0, 3), (2, 1) $2x - y = -4$
                                   $-x + y = 3$

2. (1, 1), (-5, 0), (0, -5) $2x + 5y = -25$
                                   $x - y = 5$

3. (6, 7), (-7, -6), (2, 5) $2x - 3y = 4$
                                   $7x + 7y = -91$

4. (¼, ½), (4, 2), (3, 1) $4x - y = ½$
                                   $x + 2y = 5/4$

5. (-3, -4), (0, 1), (4, 3) $x - 2y = 5$
                                   $y - 2x = 2$

6. (5, 4), (10, 5), (10, 2) $x - 2y + 1 = 1$
                                   $4x - 3y - 25 = 0$

211

II. Solve each system by graphing and indicate whether the system is independent, inconsistent, or dependent.

7. 2x - y = -4
    -x + y = 3

8. 2x + 5y = 0
    x - y = 0

9. x - 2y = 5
    y - 2x = 2

10. 4x - y = ½
    x + 2y = 5/4

11. 2x - 3y = -5
    3x + 2y = -1

12. 2x - y = 1
    ½x + ¼y = 0

13. 2x - y = 4
    4x - 2y = 8

14. 2x - y = 4
    12x - 6y = 1

15. 4x - y = 1
    x - ¼y = ¼

16. 3x - 5y = 6
    -3x + 5y = -6

17. x - 2y = 7
    -(1/3)x+(2/3)y=-(7/3)

18. 2x - y = 8
    y + x = 1

19. 3x - 5y = 6
    2x + y = 1

20. ½x - ½y = -1
    3x + 4y = 1

◆◆◆◆◆

SOLUTIONS TO EXERCISES

I. Determine which of the given points is a solution to each system of equations.

1. (-1, 2), (0, 3), (2, 1)

    2x - y = -4
    -x + y = 3

    Replace x with -1 and y with 2 in each equation of the system
    2x-y=2(-1)-(2)=-4(true) and -x+y=-(-1)+(2)=3(true).
    Since (-1,2) makes both equations true,
    therefore (-1, 2) is a solution to the system.
    (0, 3) and (2, 1) do not satisfy the system.

2. (1, 1), (-5, 0), (0, -5)

    2x + 5y = -25
    x - y = 5

    Replace x = 0 and y = -5 in each equation of the system.
    2x+5y=2(0)+5(-5)=-25 (true) and x-y=(0)-(-5)=5 (true).
    Therefore (0, -5) is a solution to the system.
    (1, 1) and (-5, 0) do not satisfy the system.

3. (6, 7), (-7, -6), (2, 5)

    2x - 3y = 4
    7x + 7y = -91

    Replace x = -7 and y = -6 in each equation of the system
    2x-3y=2(-7)-3(-6)=4 (true) and 7x+7y=7(-7)+7(-6)=-91(true).
    Therefore (-7, -6) is a solution to the system.
    (6, 7) and (2, 5) do not satisfy the system.

212

4. $(\frac{1}{4}, \frac{1}{2})$, $(4, 2)$, $(3, 1)$ $\qquad\qquad\qquad\qquad$ $4x - y = \frac{1}{2}$
$\qquad\qquad\qquad\qquad\qquad\qquad\qquad\qquad$ $x + 2y = 5/4$

Replace $x = \frac{1}{4}$ and $y = \frac{1}{2}$ in each equation of the system
$4x-y=4(\frac{1}{4})-(\frac{1}{2})=\frac{1}{2}$ (true) and $x+2y=(\frac{1}{4})+2(\frac{1}{2})=5/4$ (true).
Therefore $(\frac{1}{4}, \frac{1}{2})$ is a solution to the system.
$(4, 2)$ and $(3, 1)$ do not satisfy the system.

5. $(-3, -4)$, $(0, 1)$, $(4, 3)$ $\qquad\qquad\qquad\qquad$ $x - 2y = 5$
$\qquad\qquad\qquad\qquad\qquad\qquad\qquad\qquad$ $y - 2x = 2$

Replace $x = -3$ and $y = -4$ in each equation of the system
$x-2y=(-3)-2(-4)=5$ (true) and $y-2x=(-4)-2(-3)=2$ (true).
Therefore $(-3, -4)$ is a solution to the system.
$(0, 1)$ and $(4, 3)$ do not satisfy the system.

6. $(5, 4)$, $(10, 5)$, $(10, 2)$ $\qquad\qquad\qquad\qquad$ $x - 2y + 1 = 1$
$\qquad\qquad\qquad\qquad\qquad\qquad\qquad\qquad$ $4x - 3y - 25 = 0$

Replace $x = 10$ and $y = 5$ in each equation of the system
$x-2y+1=(10)-2(5)+1=1$ (true) and $4x-3y-25=4(10)-3(5)-25=0$ (true).
Therefore $(10, 5)$ is a solution to the system.
$(10, 2)$ and $(5, 4)$ do not satisfy the system.

II. Solve each system by graphing and indicate whether the system is
independent, inconsistent, or dependent.

(Hint: Express the system in slope-intercept form for easy graphing)

7. $\qquad$ $2x - y = -4$ $\quad$ -- (1)
$\qquad$ $-x + y = 3$ $\qquad$ -- (2)

Express the system in slope-intercept form
$y = 2x + 4$, $m = 2$ and y-intercept $(0, 4)$ -- from eq.(1)
$y = x + 3$, $m = 1$ and y-intercept $(0, 3)$ -- from eq.(2)
This is an independent system.
The point of intersection is $(-1, 2)$.

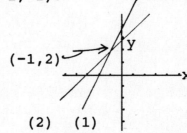

213

8.   2x + 5y = 0 -- (1)
      x - y = 0   -- (2)
      Express the system in slope-intercept form
      y = -(2/5)x, m = -2/5 and y-intercept (0, 0) -- from eq.(1)
      y = x, m = 1 and y-intercept (0, 0) -- from eq.(2)
      This is an independent system.
      The point of intersection is (0, 0).

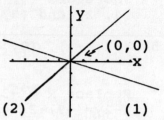

9.   x - 2y = 5 -- (1)
      y - 2x = 2 -- (2)

      Express the system in slope-intercept form
      y = ½x - 5/2, m = ½ and y-intercept (0, -5/2) -- from eq.(1)
      y = 2x + 2 where m = 2 and (0, 2) -- from eq.(2)
      This is an independent system.
      The point of intersection is (-3, -4).

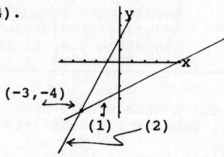

10.   4x - y = ½   -- (1)
      x + 2y = 5/4   -- (2)

      Express the system in slope-intercept form
      y = 4x - ½, m = 4 and y-intercept (0, -½) -- from eq.(1)
      y = -½x + 5/8, m = -½ and y-intercept (0, 5/8) -- from eq.(2)
      This is an independent system.
      The point of intersection is (¼, ½).

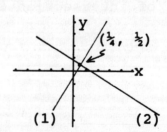

214

11. $2x - 3y = -5$ -- (1)
     $3x + 2y = -1$ -- (2)

     Express the system in slope-intercept form
     $y = (2/3)x + 5/3$, m = 2/3 and y-intercept (0, 5/3) -- from eq.(1)
     $y = -(3/2)x - \frac{1}{2}$, m = -3/2 and y-intercept (0, $-\frac{1}{2}$) -- from eq.(2)
     This is an independent system.
     The point of intersection is (-1, 1).

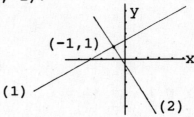

12. $2x - y = 1$ -- (1)
     $\frac{1}{2}x + \frac{1}{4}y = 0$ -- (2)

     Express the system in slope-intercept form
     $y = 2x - 1$, m = 2 and y-intercept (0, -1) -- from eq.(1)
     $\frac{1}{2}x + \frac{1}{4}y = 0$ -- multiply both sides by 4 -- from eq.(2)
     $y = -2x$, m = -2 and y-intercept (0, 0)
     This is an independent system.
     The point of intersection is ($\frac{1}{4}$, $-\frac{1}{2}$).

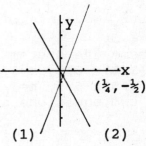

13. $2x - y = 4$ -- (1)
     $4x - 2y = 8$ -- (2)

     Express the system in slope-intercept form
     $y = 2x - 4$, m = 2 and y-intercept (0,-4) -- from eq.(1)
     $y = 2x - 4$, m = 2 and y-intercept (0,-4) -- from eq.(2)
     The two equations are identical. (Dependent)

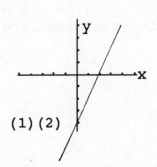

14.   2x - y = 4   -- (1)
        12x - 6y = 1   -- (2)
        Express the system in slope-intercept form
        y = 2x - 4, m = 2 and y-intercept (0, -4)   -- from eq.(1)
        y = 2x - 1/6, m = 2 and y-intercept (0,-1/6) -- from eq.(2)
        The two lines are parallel. (Inconsistent)

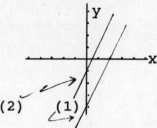

15.   4x - y = 1   -- (1)
        x - ¼y = ¼   -- (2)
        Express the system in slope-intercept form
        y = 4x - 1, m = 4 and y-intercept (0,-1) -- from eq. (1)
        y = 4x - 1, m = 4 and y-intercept (0,-1) -- from eq. (2)
        The two equations are identical. (Dependent)

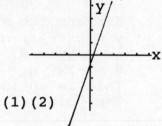

16.   3x - 5y = 6   -- (1)
        -3x + 5y = -6   -- (2)
        Express the system in slope-intercept form
        y =(3/5)x-6/5, m=3/5 and y-intercept (0, -6/5) -- from eq.(1)
        y =(3/5)x-6/5, m=3/5 and y-intercept (0,-6/5) -- from eq.(2)
        The two equations are identical. (Dependent)

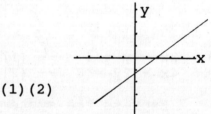

17.   x - 2y = 7   -- (1)
        -(1/3)x + (2/3)y = -(7/3) -- (2)
        Express the system in slope-intercept form
        y = ½x - 7/2, m = ½ and y-intercept (0,-7/2) -- from eq.(1)
        y = ½x - 7/2, m = ½ and y-intercept (0,-7/2) -- from eq.(2)
        The two equations are identical. (Dependent)

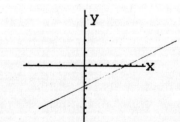

18.   2x - y = 8 -- (1)
       y + x = 1 -- (2)
       Express the system in slope-intercept form
       y = 2x - 8, m = 2 and y-intercept (0, -8) -- from eq.(1)
       y = -x + 1, m = -1 and y-intercept (0, 1) -- from eq.(2)
       This is an independent system.
       The point of intersection is (3, -2).

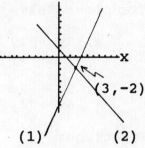

19.   3x - 5y = -5 -- (1)
       2x + y = 1   -- (2)
       Express the system in slope-intercept form
       y = (3/5)x + 1, m = 3/5 and y-intercept (0, 1) -- from eq.(1)
       y = -2x + 1, m = -2 and y-intercept (0, 1)   -- from eq.(2)
       This is an independent system.
       The point of intersection is (0, 1).

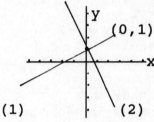

20.   ½x - ½y = -1 -- (1)
       3x + 4y = 1 -- (2)
       Express the system in slope-intercept form
       y = x + 2, m = 1 and y-intercept (0, 2) -- from eq.(1)
       y = -(3/4)x + ¼, m = -3/4 and y-intercept (0, ¼) -- from eq.(2)
       This is an independent system.
       The point of intersection is (-1, 1).

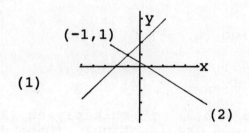

♦♦♦♦♦

217

9.2 SOLVING SYSTEMS OF LINEAR EQUATIONS BY SUBSTITUTION

OBJECTIVES:

*(A) To solve a system of equations by substitution.

Examples: Solve by substitution.

a. $3x - 2y = 5$ -- (1)
   $-x + y = -2$ -- (2)

b. $\frac{1}{4}x - \frac{1}{2}y = \frac{1}{4}$ -- (1)
   $3x + 4y = -2$ -- (2)

c. $3x + 2y = 5$ -- (1)
   $6x + 4y = 10$ -- (2)

Solutions:

a. $3x - 2y = 5$ -- (1)
   $-x + y = -2$ -- (2)

   $-x + y = -2$ -- from eq.(2)
   $y = x - 2$ -- express y in terms of x
   $3x - 2[x - 2] = 5$ -- substitute y = [x - 2] into eq.(1)
   $3x - 2x + 4 = 5$ -- simplify
   $x = 1$
   $-(1) + y = -2$ -- substitute x = 1 into eq.(2) to find y
   $y = -1$
   The solution is (1, -1).

b. $\frac{1}{4}x - \frac{1}{2}y = \frac{1}{4}$ -- (1)
   $3x + 4y = -2$ -- (2)

   $\frac{1}{4}x - \frac{1}{2}y = \frac{1}{4}$ -- from eq.(1)
   $x - 2y = 1$ -- multiply both sides by 4
   $x = 2y + 1$ -- express x in terms of y
   $3[2y + 1] + 4y = -2$ -- substitute x = [2y + 1] into eq.(2)
   $6y + 3 + 4y = -2$ -- simplify
   $10y = -5$
   $y = -\frac{1}{2}$ -- substitute y = -½ into eq.(1) to find x
   $\frac{1}{4}x - \frac{1}{2}(-\frac{1}{2}) = \frac{1}{4}$ -- multiply both sides by 4
   $x + 1 = 1$
   $x = 0$
   The solution (0, -½).

c. $3x + 2y = 5$ -- (1)
   $6x + 4y = 10$ -- (2)

   $2(3x + 2y) = 2(5)$ -- multiply eq.(1) by 2 on both sides
   Two equations are identical. (Dependent)

◆◆◆◆◆

◆◆◆◆◆

**EXERCISES

I. Solve each system by the substitution method.

1. $4y - 3x = -1$
   $2y + x = -\frac{1}{2}$

2. $x - 2y = -4$
   $5x - 6y = -16$

3. $12x + y = -8$
   $y - 5x = 9$

4. $-2x + 3y = 5$
   $x - 8y = 4$

5. $x + \frac{1}{2}y = \frac{1}{4}$
   $2y - 1 = x$

6. $2x - 3y = -1$
   $5x - 4y = 1$

7. $3x - 4y = 1$
   $5x - 6y = 3$

8. $x - 5y = -5$
   $5x - y = -25$

9. $x + y = 3$
   $\frac{1}{4}x + y = 0$

10. $4x + 5y = -1$
    $9x + y = 8$

II. Solve each system by the substitution method and identify each system as independent, dependent, or inconsistent.

11. $4y - 3x = -1$
    $8y - 6x = -2$

12. $x + y = 3$
    $5x + 5y = 1$

13. $x + y = 2$
    $\frac{1}{2}x + \frac{1}{2}y = \frac{1}{4}$

14. $2x - 3y = 15$
    $\frac{1}{4}x - (1/5)y = 1$

15. $6x + 7y = 5$
    $-y + x = 3$

16. $2x - 7y = 4$
    $x + y = -7$

17. $2x + y = 7$
    $x + y = -1$

18. $8x - 5y = 0$
    $2x - y = 2$

19. $2x + \frac{1}{2}y = -1$
    $-7x - 3y = 1$

20. $\frac{1}{4}x + \frac{1}{2}y = \frac{1}{2}$
    $x + 2y = 2$

◆◆◆◆◆

**SOLUTIONS TO EXERCISES

I. Solve each system by the substitution method.

1. $4y - 3x = -1$ -- (1)
    $2y + x = -\frac{1}{2}$ -- (2)

    $x = -\frac{1}{2} - 2y$ -- express x in terms of y (from eq.(2))
    $4y - 3[-\frac{1}{2} -2y] = -1$ -- substitute $x = [-\frac{1}{2} - 2y]$ into eq.(1)
    $4y + 3/2 + 6y = -1$ -- simplify
    $y = -\frac{1}{4}$, $x = -\frac{1}{2} - 2(-\frac{1}{4}) = 0$
    $x = 0$, $y = -\frac{1}{4}$
    $(0, -\frac{1}{4})$ is the solution.

2. $x - 2y = -4$ -- (1)
   $5x - 6y = -16$ -- (2)

   $x = -4 + 2y$ -- from eq.(1)
   $5[-4 + 2y] - 6y = -16$ -- substitute x = [-4 + 2y] into eq.(2)
   $-20 + 10y - 6y = -16$
   $y = 1,$ $x = -4 + 2y = -2$
   $(-2, 1)$ is the solution.

3. $12x + y = -8$ -- (1)
   $y - 5x = 9$ -- (2)

   $y = -8 - 12x$ -- from eq.(1)
   $[-8 -12x] - 5x = 9$ -- substitute y = [-8 -12x] into eq.(2)
   $-17x = 17$
   $x = -1,$ $y = -8 - 12(-1) = 4$
   $(-1, 4)$ is the solution.

4. $-2x + 3y = 5$ -- (1)
   $x - 8y = 4$ -- (2)

   $x = 4 + 8y$ -- from eq.(2)
   $-2[4 + 8y] + 3y = 5$ -- substitute x = [4 + 8y] into eq.(1)
   $-13y = 13$
   $y = -1,$ $x = 4 + 8(-1) = -4$
   $(-4, -1)$ is the solution.

5. $x + \frac{1}{2}y = \frac{1}{4}$ -- (1)
   $2y - 1 = x$ -- (2)

   $x = 2y - 1$ -- from eq.(2)
   $[2y - 1] + \frac{1}{2}y = \frac{1}{4}$ -- substitute x = [2y -1] into eq.(1)
   $(5/2)y = 5/4$
   $y = \frac{1}{2}$ $x = 2(\frac{1}{2}) - 1 = 0$
   $(0, \frac{1}{2})$ is the solution.

6. $2x - 3y = -1$ -- (1)
   $5x - 4y = 1$ -- (2)

   $x = -\frac{1}{2} + (3/2)y$ -- from eq.(1)
   $5[-\frac{1}{2}+ 3/2y] - 4y = 1$ -- substitute x = [-½+(3/2)y] into eq.(2)
   $(7/2)y = 7/2$
   $y = 1,$ $x = -\frac{1}{2} + (3/2)(1) = 1$
   $(1, 1)$ is the solution.

7. $3x - 4y = 1$ -- (1)
   $5x - 6y = 3$ -- (2)

   $x = (1/3)[1 + 4y]$ -- from eq.(1)
   $5(1/3)[1 + 4y] - 6y = 3$ -- substitute x = (1/3)[1 + 4y] into eq.(2)
   $y = 2,$ $x = (1/3)[1 + 4(2)] = 3$
   $(3, 2)$ is the solution.

8. x - 5y = -5   -- (1)
    5x - y = -25   -- (2)

    x = 5y - 5   -- from eq.(1)
    5[5y - 5] - y = -25   -- substitute x = [5y-5] into eq.(2)
    y = 0,   x = 5(0) - 5 = -5
    (-5, 0) is the solution.

9. x + y = 3   -- (1)
    ¼x + y = 0   -- (2)

    x = 3 - y   -- from eq.(1)
    ¼[3 - y] + y = 0   -- substitute x = [3 - y] into eq.(2)
    y = -1,   x = 3 - (-1) = 4
    (4, -1) is the solution.

10. 4x + 5y = -1   -- (1)
    9x + y = 8   -- (2)

    y = 8 - 9x   -- from eq.(2)
    4x + 5[8 - 9x] = -1   -- substitute y = [8 - 9x] into eq.(1)
    -41x = -41
    x = 1, y = 8 - 9(1) = -1
    (1, -1) is the solution.

◆◆◆◆◆

II. Solve each system by the substitution method and identify each
    system as independent, dependent, or inconsistent.

11. 4y - 3x = -1   12. x + y = 3   13. x + y = 2
     8y - 6x = -2   5x + 5y = 1   ½x + ½y = ¼
     m1 = m2 = -1   m1 = m2 = -1
        (Dependent)   (Inconsistent)   (Inconsistent)
The two equations are identical.

14. 2x - 3y = 15   -- (1)
     ¼x - (1/5)y = 1 -- (2)

     x = ½(3y + 15)   -- from eq.(1)
     ¼[½(3y + 15)]- (1/5)y = 1   -- substitute x = ½[3y + 15] into eq.(2)
     5(3y + 15) - 8y = 40   -- multiply both sides by 40 (LCD)
     7y = -35
     y = -5, x = ½[3(5) + 15] = 0
     (0, -5) is the solution.

15. 6x + 7y = 5   --(1)   16. 2x - 7y = 4 -- (1)
     -y + x = 3   --(2)   x + y = -7   -- (2)

     y = x - 3 -- from eq.(2)   y = -7 - x -- from eq.(2)
     6x + 7[x - 3] = 5   2x - 7[-7 -x] = 4
     x = 2, y = 2 - 3 = -1   x = -5, y = -7-(-5) = -2
     (2, -1) is the solution.   (-5, -2) is the solution.

221

17. $2x + y = 7$ -- (1)
     $x + y = -1$ -- (2)
     $x = -1 - y$ -- from eq.(2)
     $2[-1 - y] + y = 7$ substitute $x = [-1 - y]$ into eq.(2)
     $y = -9$, $x = -1 -(-9) = 8$
     (8, -9) is the solution.

18. $8x - 5y = 0$ -- (1)
     $2x - y = 2$ -- (2)
     $y = 2x - 2$ -- from eq.(2)
     $8x - 5[2x - 2] = 0$ substitute $y = [2x - 2]$ into eq.(1)
     $x = 5$, $y = 2(5) - 2 = 8$
     (5, 8) is the solution.

19. $2x + \frac{1}{2}y = -1$ -- (1)
     $-7x - 3y = 1$ -- (2)
     $x = -\frac{1}{2} - \frac{1}{4}y$ -- from eq.(1)
     $-7[-\frac{1}{2} - \frac{1}{4}y] - 3y = 1$ -- substitute $x = [-\frac{1}{2} - \frac{1}{4}y]$ into eq.(2)
     $14 + 7y - 12y = 4$ -- simplify
     $y = 2$, $x = -\frac{1}{2} - \frac{1}{4}(2) = -1$ -- multiply both sides by 4 (LCD)
     (-1, 2) is the solution.

20. $\frac{1}{4}x + \frac{1}{2}y = \frac{1}{2}$
     $x + 2y = 2$
     The two lines are identical.
     (Dependent)

♦♦♦♦♦

9.3 THE ADDITION METHOD

OBJECTIVES:

*(A) To solve a system of equations by addition.

** REVIEW

*(A) To solve the system of equations by addition, remember the
     following:

  Step 1. The two equations must be in the same form.
            $a_1x + b_1y = c_1$ -- (1)
            $a_2x + b_2y = c_2$ -- (2)

  Step 2. Multiply if necessary, so that the coefficients of either x or
          y are numerically equal but opposite in sign.

  Step 3. Add the new equations to eliminate one variable.

  Step 4. Solve the equation from Step 3.

  Step 5. Substitute the value of Step 4 into either of the given
          equations and solve for the other variable.

Examples:

Solve the system by addition.

    a. $2x - y = -4$ -- (1)   b. $2x - 3y = 9$ -- (1)
       $-2x + 3y = 8$ -- (2) $4x - y = 3$ -- (2)

Solutions:
    a. $2x - y = -4$ -- (1)
        $-2x + 3y = 8$ -- (2)

The two equations are in the same form   -- step 1
The coefficients for x are 2 and -2   -- step 2
Add the two equations so that one variable
will be eliminated   -- step 3

$$\begin{array}{r} 2x - y = -4 \\ +)\ -2x + 3y = 8 \\ \hline \end{array}$$

            $2y = 4$ -- eliminate x
             $y = 2$ -- substitute $y = 2$ into eq.(1) or eq.(2)

    $2x - (2) = -4$ -- from eq.(1)
    $x = -1$ * Check: Let x=-1 and y=2
    The solution is (-1, 2). $2(-1)-(2)=-4$ correct -- (1)
                             $-2(-1)+3(2)=8$ correct -- (2)

    b. $2x - 3y = 9$ -- (1)
       $4x - y = 3$ -- (2)

The two equations are in the same form -- step 1
The coefficients for x are 2 and 4   -- step 2
Therefore multiply eq.(1) by -2
 $-4x + 6y = -18$ -- (3)
Add eq.(2) to eq.(3) so that one variable x
will be eliminated -- step 3

$$\begin{array}{r} -4x + 6y = -18 \ \text{-- (3)} \\ +)\ 4x - y = 3 \ \ \ \ \ \text{-- (2)} \\ \hline \end{array}$$

            $5y = -15$

$y = -3$ -- substitute $y = -3$ into eq.(1) or eq.(2)
$2x - 3(-3) = 9$ -- from eq.(1)
$x = 0$ * Check: Let x=0 and y=-3
The solution is (0, -3).

                    $2 \cdot 0 - 3(-3) = 9$ correct --(1)
                    $4 \cdot 0 - (-3) = 3$ correct --(2)

♦♦♦♦♦

I. Solve each system by the addition method.

1. $2x - y = -2$
$6x + 8y = 9$

2. $x + \frac{1}{2}y = 2$
$-3x + y = 9$

3. $x + 7y = -1$
$3x + 25y = 1$

4. $2x - 3y = 6$
$5x + 3y = -6$

5. $9x + 4y = 11$
$4x + 3y = 0$

6. $7x + 6y = -1$
$6x + 7y = 1$

7. $12x + 5y = -6$
$2x + 3y = -1$

8. $-x + y = 5$
$3x + 4y = -1$

9. $3x - 2y = -1$
$2x - 3y = 11$

10. $2x + y = 18$
$4x - y = 6$

II. Solve each system and state whether the system is independent, dependent, or inconsistent.

11. $9x - y = 10$
$18x - 2y = 5$

12. $x + y = -5$
$-2x - 6y = 22$

13. $3x + 11y = \frac{1}{2}$
$30x + 110y = 5$

14. $x - 3y = -3$
$3x - y = -1$

15. $\frac{1}{2}x - \frac{1}{4}y = \frac{3}{4}$
$2x - y = 3$

16. $12x - 14y = 7$
$6x - 7y = 14$

17. $y - \frac{1}{4} = x$
$2x + \frac{3}{4} = 2y$

18. $x - 2y = 1$
$x - 2y = 2$

19. $-3x = 6y + 6$
$7x = 2y - 6$

20. $5x + 4y = 10$
$20x + 16y = 5$

♦♦♦♦♦

**SOLUTIONS TO EXERCISES

I. Solve each system by the addition method.
   (Check by substitution in the original equation.)

1. $2x - y = -2$ --(1)
$6x + 8y = 9$ -- (2)

$(-3)(2x - y) = (-3)(-2)$ -- multiply eq.(1) by -3

$-6x + 3y = 6$ -- simplify the above equation
+) $6x + 8y = 9$ -- from eq.(2)

$11y = 15$
$y = 15/11, \quad x = -(7/22)$
The solution is $(-7/22, 15/11)$.

2. $x + \frac{1}{2}y = 2$ -- (1)
     $-3x + y = 9$ -- (2)

      $3(x + \frac{1}{2}y) = 3(2)$ -- multiply eq.(1) by 3

      $3x + (3/2)y = 6$ -- simplify the above equation
   +) $-3x + y = 9$ -- from eq.(2)
   ─────────────────────────
          $(5/2)y = 15$
      $y = 6, x = -1$
      The solution is $(-1, 6)$.

3. $x + 7y = -1$ -- (1)
     $3x + 25y = 1$ -- (2)

      $(-3)(x + 7y) = (-3)(-1)$ -- multiply eq.(1) by -3

      $-3x - 21y = 3$ -- simplify the above equation
   +) $3x + 25y = 1$ -- from eq.(2)
   ─────────────────────────
          $4y = 4$

      $y = 1, x = -1 - 7(1) = -8$
      The solution is $(-8, 1)$.

4. $2x - 3y = 6$ -- (1)
     $5x + 3y = -6$ -- (2)

      $7x = 0$ -- add eq.(1) to eq.(2)
      $x = 0, y = -2$
      The solution is $(0, -2)$.

5. $9x + 4y = 11$ -- (1)
     $4x + 3y = 0$ -- (2)

      $(-3)(9x + 4y) = (-3)(11)$ -- multiply eq.(1) by -3
      $-27x - 12y = -33$ -- (3)
      $4(4x + 3y) = 0$ -- multiply eq.(2) by 4
      $16x + 12y = 0$ -- (4)
      Add eq.(3) to eq.(4)

      $-27x - 12y = -33$ -- from eq.(3)
   +) $16x + 12y = 0$ -- from eq.(4)
   ─────────────────────────
      $-11x = -33$

      $x = 3, y = -4$
      The solution is $(3, -4)$.

6.　$7x + 6y = -1$ -- (1)
　　$6x + 7y = 1$ -- (2)

　　$(-6)(7x + 6y) = (-6)(-1)$ -- multiply eq.(1) by -6
　　$-42x - 36y = 6$ -- (3)
　　$7(6x + 7y) = 7(1)$ -- multiply eq.(2) by 7
　　$42x + 49y = 7$ -- (4)
　　Add eq.(3) to eq.(4)

　　$-42x - 36y = 6$ -- from eq.(3)
+) $42x + 49y = 7$ -- from eq.(4)
　　―――――――――――
　　　　$13y = 13$

　　$y = 1, \quad x = -1$
　　The solution is (-1, 1).

7.　$12x + 5y = -6$ -- (1)
　　$2x + 3y = -1$ -- (2)

　　$(-6)(2x + 3y) = (-6)(-1)$ -- multiply eq.(2) by -6
　　$-12x - 18y = 6$ -- (3)
　　Add eq.(1) and eq.(3)

　　$-12x - 18y = 6$ -- from eq.(3)
+) $12x + 5y = -6$ -- from eq.(1)
　　―――――――――――
　　　　$-13y = 0$

　　$y = 0, \quad x = -\frac{1}{2}$
　　The solution is $(-\frac{1}{2}, 0)$.

8.　$-x + y = 5$ -- (1)
　　$3x + 4y = -1$ -- (2)

　　$(3)(-x + y) = (3)(5)$ -- multiply eq.(1) by 3
　　$-3x + 3y = 15$ -- (3)
　　Add eq.(2) to eq.(3)

　　$-3x + 3y = 15$ -- from eq.(3)
+) $3x + 4y = -1$ -- from eq.(2)
　　―――――――――――
　　　　$7y = 14$

　　$y = 2, \quad x = -3$
　　The solution is (-3, 2).

9.   3x - 2y = -1   -- (1)
      2x - 3y = 11   -- (2)

      (-3)(3x -2y) = (-3)(-1) -- multiply eq.(1) by -3
      -9x + 6y = 3 -- (3)
      2(2x - 3y) = 2(11)   -- multiply eq.(2) by 2
      4x - 6y = 22 -- (4)
      Add eq.(3) and eq.(4)

      -9x + 6y = 3 -- from eq.(3)
   +) 4x - 6y = 22 -- from eq.(4)
      _____

      -5x = 25

      x = -5, y = -7
      The solution is (-5, -7).

10. 2x + y = 18   -- (1)
     4x - y = 6   -- (2)

     6x = 24   -- add eq.(1) to eq.(2)
     x = 4,   y = 10
     The solution is (4, 10).

 II. Solve each system and state whether the system is independent,
     dependent, or inconsistent.

11. 9x - y = 10   -- (1)   12.   x + y = -5   -- (1)
     18x - 2y = 5 -- (2)   -2x -6y = 22   -- (2)

     y = 9x - 10   -- from eq.(1)   2x + 2y = -10 -- eq.(1)x2
     y = 9x -(5/2) -- from eq.(2)   +) -2x - 6y = 22 -- from eq.(2)
     m1 = m2 = 9   _____
     This is an inconsistent system.   -4y = 12 implies y = -3
                                          The solution is (-2, -3).

13. 3x + 11y = ½   -- (1)
     30x + 110y = 5 -- (2)

     The two equations are identical.
     (Dependent)

14. x - 3y = -3 -- (1)
     3x - y = -1 -- (2)

     (-3)(x - 3y) = (-3)   -- multiply eq.(1) by -3

     -3x + 9y = 9 -- simplify the above equation
   +) 3x - y = -1 -- from eq.(2)
      _____

          8y = 8

     y = 1, x = 0
     The solution is (0, 1).

227

15. $\frac{1}{2}x - \frac{1}{4}y = \frac{3}{4}$ -- (1)
    $2x - y = 3$ -- (2)

    Two equations are identical.
    (Dependent)

16. $12x - 14y = 7$ -- (1)
    $6x - 7y = 14$ -- (2)

    $y = 6/7x - \frac{1}{2}$ -- from eq.(1)
    $y = 6/7x - 2$ -- from eq.(2)
    (Inconsistent)

17. $y - \frac{1}{4} = x$ -- (1)
    $2x + \frac{3}{4} = 2y$ -- (2)

    $y = x + \frac{1}{4}$ -- from eq.(1)
    $y = x + 3/8$ -- from eq.(2)
    $m1 = m2 = 1$
    (Inconsistent)

18. $x - 2y = 1$ -- (1)
    $x - 2y = 2$ -- (2)

    $y = \frac{1}{2}x - \frac{1}{2}$ -- from eq.(1)
    $y = \frac{1}{2}x - 1$ -- from eq.(2)
    $m1 = m2 = \frac{1}{2}$
    (Inconsistent)

19. $-3x = 6y + 6$ -- (1)
    $7x = 2y - 6$ -- (2)

    Multiply eq.(2) by -3
    $-21x = -6y + 18$ -- (3)
    Add eq.(1) and eq.(3)
    $-24x = 24$
    $x = -1$, $y = -\frac{1}{2}$
    The solution is $(-1, -\frac{1}{2})$.

20. $5x + 4y = 10$ -- (1)
    $20x + 16y = 5$ -- (2)

    $y = -(5/4)x + 5/2$ - from eq.(1)
    $y = -(5/4)x + 5/16$ - from eq.(2)
    $m1 = m2 = -(5/4)$
    (Inconsistent)

♦♦♦♦♦

9.4 LINEAR INEQUALITIES IN TWO VARIABLES

OBJECTIVES:

*(A) To define a linear inequality in two variables.
*(B) To solve a linear inequality in two variables.
*(C) To graph a linear inequality in two variables.

** REVIEW

*(A)
 A linear inequality in two variables has the following forms.

$$ax + by \geq c \qquad ax + by \leq c$$
$$ax + by > c \qquad ax + by < c,$$

where a,b, and c are real, and a and b both not zero.

Examples:
>   a. 3x > 2y + 5   -- subtract 2y from both sides
>   3x - 2y > 5   -- in the form of ax - by > c
>
>   b. 5 < 2y - x   -- rearrange the terms
>   -x + 2y > 5   -- in the form of ax + by > c
>
>   c. y ≤ 5 + 7x   -- rearrange the terms
>   7x - y ≥ -5   -- in the form of ax - by ≥ c

*(B) To solve a linear inequality in two variables is to find the
     ordered pairs that satisfies the inequality.

Examples:
 Determine whether or not each of the points satisfies the inequality.

>   a. x - 5y ≥ 10   (3,0), (4,-1), (10,-1)
>
>   b. 2x + y ≤ -1   (10,0), (0,-5), (-5,-5)

Solutions:
>   a. x - 5y ≥ 10   (3,0),(4,-1),(10,-1)
>
>   Consider the ordered pair (3,0), replace x with 3 and y with 0,
>   x - 5y = (3) - 5(0) = 3 ≥ 10   incorrect
>
>   Consider the ordered pair (4,-1), replace x with 4 and y with -1
>   x - 5y = (4) - 5(-1) = 9 ≥ 10   incorrect
>
>   Consider the ordered pair (10,-1), replace x with 10 and y with -1
>   x - 5y = (10) - 5(-1) = 15 ≥ 10   correct
>   (3,0) and (4,-1) do not satisfy the inequality but (10,-1) does.
>
>   b. 2x + y ≤ -1   (10, 0),(0, -5),(-5, -5)
>
>   Consider the ordered pair (10,0), replace x with 10 and y with 0,
>
>   2x + y = 2(10) + (0) = 20 ≤ -1   incorrect
>   Consider the ordered pair (0,-5), replace x with 0 and y with -5
>
>   2x + y = 2(0) + (-5) = -5 ≤ -1   correct
>   Consider the ordered pair (-5,-5), replace x with -5 and y with -5
>
>   2x + y = 2(-5) + (-5) = -15 ≤ -1   correct
>   (10,0) does not satisfy the inequality but (0,-5) and (-5,-5) do.

*(C) To graph a linear inequality is to locate all the points on the
     rectangular coordinate system that satisfy the inequality.
   ● Choose any point not on the line as a test point.

Examples: Graph the inequalities.

>   a. y - 3x > 5   b. 3x - 2y ≤ 7   c. x ≤ -2

Solutions:

    a. $y - 3x > 5$ -- try to graph $y - 3x = 5$
       $y = 3x + 5$ -- in slope-intercept form
       $m = 3$ and y-intercept $(0, 5)$.

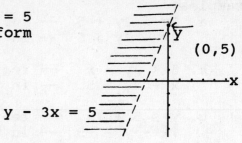

The set of points satisfying the inequality $y - 3x > 5$
is the region **to the left** of the line $y - 3x = 5$.
The line $y - 3x = 5$ is not included (dashed line).

    b. $3x - 2y \leq 7$
      Graph $3x - 2y = 7$
      $y = (3/2)x - 7/2$ -- in slope-intercept form
      $m = 3/2$ and y-intercept $(0, -7/2)$.

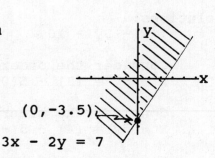

The set of points satisfying the inequality $3x - 2y \leq 7$
is the region **above** the line $3x - 2y = 7$.
The line $3x - 2y = 7$ is included (solid line).

    c. $x \leq -2$
      The line $x = -2$ is the vertical line

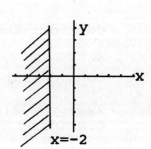

The set of points satisfying the inequality $x \leq -2$
is the region **to the left** of the line $x = -2$.
The line $x = -2$ is included (solid line).

◆◆◆◆◆

I. True or False?

1. The graph of $x \leq -5$ is a horizontal line 5 units below the y-axis.

2. The point $(0, -2)$ satisfies the inequality $2x - 3y \leq 7$.

3. The graph of the inequality $2x \geq 2y + 3$ is the region above the line $2x = 2y + 3$ but the line is not included.

4. The graph of the inequality $x \leq -3 + 2y$ is the region above the line $x = -3 + 2y$ but the line is included.

5. The graph $y > -5$ is the region above the line $y = -5$.

6. The point $(0, 5)$ satisfies the inequality $6x < 2y - 10$.

II. Determine which of these points satisfies the given inequality.

7. $x > 2y + 5$ $(0, 0)$, $(5, 0)$, $(8, 1)$
8. $y \leq -3x + 1$ $(1, 0)$, $(2, 2)$, $(3, -1)$
9. $y > -x$ $(2, -1)$, $(-3, 1)$, $(-1,-1)$
10. $y < 2x + 5$ $(0, -5)$, $(-1, 1)$, $(2, 1)$

III. Graph each inequality.

11. $x \leq -5$ 12. $y > 3$ 13. $x < 2y + 3$

14. $y < \frac{1}{2}x + \frac{1}{4}$ 15. $2y > 3x$ 16. $y \leq -x$

17. $3x < 2y + 5$ 18. $y \geq -2x + 1$

19. $y < -3$ 20. $x \geq 4$

◆◆◆◆◆

**SOLUTIONS TO EXERCISES

I. True or False?

1. False.   ($x \leq -5$ is the region to the left of the vertical line $x = -5$)

2. True.   ($2x - 3y = 2(0) - 3(-2) = 6 \leq 7$)

3. False. (The region represents $y \leq x - 3/2$ is the region below the line $y = x - 3/2$ and the line is included)

4. True. (graph $y = \frac{1}{2}x + 3$)

5. True.

6. False. (replace x by 0 and y by 5 in the inequality
            6(0) < 2(5) - 10 is false)

II. Determine which of these points satisfies the given inequality.

7. x > 2y + 5   (0, 0), (5, 0), (8, 1)
   8 > 2(1) is correct -- replace x = 8 and y = 1 in the inequality
   (8, 1) satisfies the inequality.
   (0, 0) and (5, 0) do not satisfy the inequality.

8. y ≤ -3x + 1   (1, 0), (2, 2), (3, -1)
   (1, 0), (2, 2), (3, -1) do not satisfy the inequality.

9. y > -x   (2, -1), (-3, 1), (-1, -1)
   -1 > -(2) is correct -- replace x = 2 and y = -1 in the inequality
   (2, -1) satisfies the inequality.
   (-3, 1) and (-1, -1) do not satisfy the inequality.

10. y < 2x + 5   (0, -5), (-1, 1), (2, 1)
    - 5 < 2(0) + 5 is correct   -- replace x = 0 and y = -5
    1 < 2(-1) + 5 is correct   -- replace x = -1 and y = 1
    1 < 2(2) + 5 is correct   -- replace x = 2 and y = 1
    (0, -5), (-1, 1), and (2, 1) satisfy the inequality.

III. Graph each inequality.

11. x ≤ -5   12. y > 3   13. x > 2y + 3

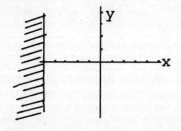   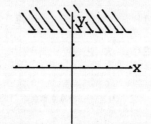   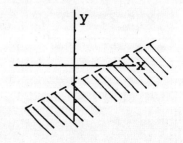

14. y < ½x + ¼   15. 2y > 3x   16. y ≤ -x

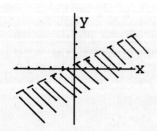   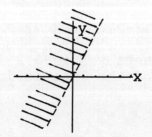   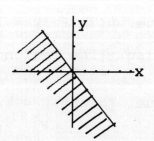

232

17. 3x < 2y + 5

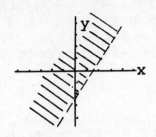

18. y ≥ -2x + 1

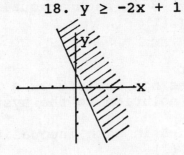

19. y < -3

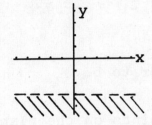

20. x ≥ 4

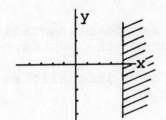

♦♦♦♦♦

9.5   SYSTEMS OF LINEAR INEQUALITIES

OBJECTIVES:

*(A) To solve a system of inequalities.
*(B) To graph a system of inequalities.

** REVIEW

*(A) To solve a system of inequalities, find a solution that
     satisfies the system of inequalities.

Examples: Determine whether or not each point is a solution to the
          system of inequalities.

$$2x - 3y > 5 \quad \text{-- eq.(1)}$$
$$x + 2y < -1 \quad \text{-- eq.(2)}$$

    a. (0, 1)   b. (-2, -3)   c. (4, -5)

Solutions:
    a. Let x = 0 and y = 1 in each inequality
       2x - 3y > 5 -- eq.(1)   x + 2y < -1 -- eq.(2)
       2(0) - 3(1) > 5   (0) + 2(1) < -1
       -3 > 5   incorrect   2 < -1   incorrect
       Since (0, 1) does not satisfy both inequalities,
       (0, 1) is not a solution to the system.

233

b. Let x = -2 and y = -3 in each inequality

2x - 3y > 5 -- eq.(1)   x + 2y < -1 -- eq.(2)
2(-2) - 3(-3) > 5   (-2) + 2(-3) < -1
-4 + 9 > 5   -2 - 6 < -1
5 > 5   incorrect   -8 < -1   correct

Since (-2, -3) does not satisfy 2x - 3y > 5,
(-2, -3) is not a solution to the system.

c. Let x = 4 and y = -5 in each inequality

2x - 3y > 5 -- eq.(1)   x + 2y < -1 -- eq.(2)
2(4) - 3(-5) > 5   (4) + 2(-5) < -1
8 + 15 > 5   4 - 10 < -1
23 > 5   correct   -6 < -1   correct

Since (4, -5) **satisfies both inequalities**,
(4, -5) is a solution to the system.

- **A point must satisfy both inequalities in order to be
  a solution of the system.**

*(B) To graph a system of inequalities, locate all points on the plane
that belong to the graphs of both inequalities in the system.
(Choose any point not on the line as a test point)

Example: Graph   y < 2x + 1
                      y ≥ x

Solution:

Graph y = 2x + 1 where m = 2 and y-intercept (0, 1).
Graph y = x where m = 1 and y-intercept (0, 0).
The region to the right of y = 2x + 1 represents y < 2x + 1.
The region to the left of y = x represents y ≥ x.
The region that represents both is the region that satisfies
both inequalities.

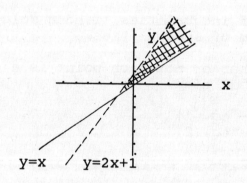

♦♦♦♦♦

234

EXERCISES

I. True or False?

1. The point (-2, 3) is a solution of the system
   $x < -8$
   $y > -5$.

2. The point ($\frac{1}{4}$, $-\frac{1}{4}$) is a solution of the system
   $4x + y > 10$
   $x - 4y < -5$.

3. The graph of the inequality $3x - 5y + 5 \leq 0$ is the region below the line $3x - 5y + 5 = 0$ and the line is included.

4. There are no solutions of the system
   $y \leq x + 3$
   $y \geq x + 3$.

5. The inequality $3x - 5 \leq y$ is equivalent to $-3x + 5y \geq -y$.

II. Determine which of the points is a solution to the given system.

6. $5x - y \leq 1$ (0, 2), (5, -5), (0, 0)
   $2x + 3y \geq 5$

7. $x > 5$ (6, -8), (5, -3), (-3, 5)
   $y < -3$

8. $2x + y \leq -1$ (0, 0), (-1, -1), (-2, 3)
   $5x + 4y > 0$

9. $x - 2y < 0$ (0, 0), (-1, 2), (3, -3)
   $x - 2y \geq 0$

10. $x > -5$ (-1, -10), (2, 5), (3, 3)
    $y \leq -10$

II. Graph each system of inequalities.

11. $y < 2x + 1$
    $x \geq 5$

12. $y - x + 5 > 0$
    $x - y + 5 < 1$

13. $2x + 3y > 5$
    $x - 5y \leq 1$

14. $2x + \frac{1}{4}y > 0$
    $y < -5$

15. $5 < x + y$
    $-2 \geq y - 2x$

16. $y < x + \frac{1}{2}$
    $x \geq 5$

17. $2x - 4y \leq -7$
    $9x + y > 0$

18. $x > -y + 1$
    $-x < y - 2$

19. $5x - 2y > 2$
    $2x + 5y \leq -5$

20. $-x + y > 5$
    $x - y > -1$

◆◆◆◆◆

SOLUTIONS TO EXERCISES

I. True or False?

1. False. (If x = -2, then -2 < -8 is incorrect)

2. False. (If x = ¼, then 4·(¼) + (-¼) = ¾ > 10 is incorrect)

3. False. (The region above the line 3x - 5y + 5 = 0)

4. False. ((0, 3), (1, 4) are solutions to the system)

5. True. (Multiply both sides by -1 and reverse the inequality)

II. Determine which of the points is a solution to the given system.

6. 5x - y ≤ 1 -- (1)   **(0, 2)**, (5, -5), (0, 0)
   2x + 3y ≥ 5 -- (2)

Consider (0,2):   Consider (5,-5) and (0,0):
Let x = 0 and y = 2   Let x = 5 and y = -5
5(0) - (2) ≤ 1 -- correct   5(5) - (-5) ≤ 1 -- incorrect
2(0) + 3(2) ≥ 5 -- correct   Let x = 0 and y = 0
(0, 2) is a solution.   2(0) + 3(0) ≥ 5 -- incorrect

7. x > 5   **(6, -8)**, (5, -3), (-3, 5)
   y < -3

Let x = 6 and y = -8   Let x = 5 and y = -3
6 > 5   -- correct   5 > 5   -- incorrect
-8 < -3 -- correct   Let x = -3 and y = 5
(6, -8) is a solution.   -3 > 5   -- incorrect

8. 2x + y ≤ -1   (0, 0), (-1, -1), **(-2, 3)**
   5x + 4y > 0

Let x = -2 and y = 3   Let x = 0 and y = 0
2(-2) + (3) ≤ -1 -- correct   2(0) + (0) ≤ -1 -- incorrect
5(-2) + 4(3) ≥ 0 -- correct   Let x = -1 and y = -1
(-2, 3) is a solution.   2(-1) + (-1) ≥ 5 -- incorrect

9. x - 2y < 0   (0, 0), (-1, 2), (3, -3)
   x - 2y ≥ 0

Let x = 3 and y = -3
3 - 2(-3) < 0 -- incorrect   Let x = 0 and y = 0
Let x = -1 and y = 2   0 - 2(0) < 0 -- incorrect
-1 - 2(2) ≥ 0 -- incorrect
no solution.

10. x > -5   (-1, -10), (2, 5), (3, 3)
    y ≤ -10

    Let x = -1 and y = -10   Let x = 2 and y = 5
    -1 > -5   -- correct   5 ≤ -10   -- incorrect
    -10 ≤ -10   -- correct   Let x = 3 and y = 3
    (-1, -10) is a solution.   3 ≤ -10   -- incorrect

II. Graph each system of inequalities.

11. y < 2x + 1 --(1)   12. y - x + 5 > 0 --(1)   13. 2x + 3y > 5 -- (1)
    x ≥ 5 -- (2)   x - y + 5 < 1 -- (2)   x - 5y ≤ 1 -- (2)

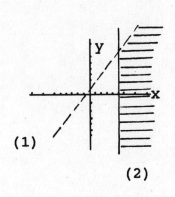

(1)

(2)

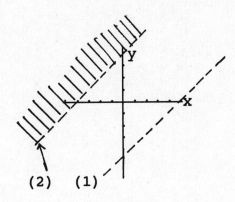

(2)   (1)

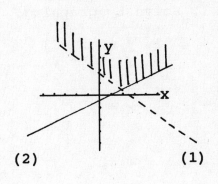

(2)   (1)

14. 2x + ¼y > 0 --(1)   15. 5 < x + y -- (1)   16. y < x + ½ -- (1)
    y < -5 -- (2)   -2 ≥ y - 2x -- (2)   x ≥ 5 -- (2)

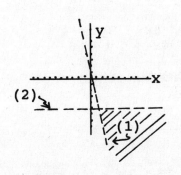

(2)

(1)

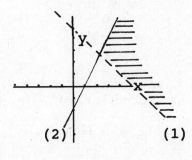

(2)   (1)

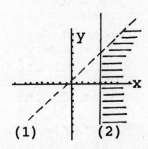

(1)   (2)

17. 2x - 4y ≤ -7 - (1)   18. x > -y + 1 -- (1)   19. 5x - 2y > 2 -- (1)
    9x + y > 0   - (2)   -x < y - 2 -- (2)   2x + 5y ≤ -5 --(2)

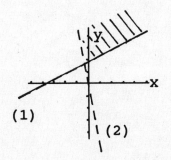

(1)

(2)

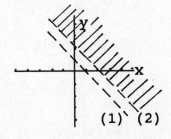

(1) (2)

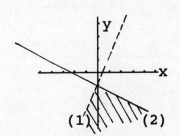

(1)   (2)

237

20. $-x + y > 5$ -- (1)
    $x - y > -1$ -- (2)

There is no solution to this system,
because there is no common region above
the line $-x+y=5$ and below the line
$x-y=-1$.

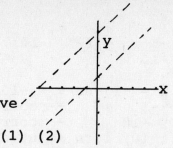

(1) (2)

♦♦♦♦♦

CHAPTER IX - TEST A

I. Solve by graphing.

1. $y = \frac{1}{2}x + 3$
   $y = 1 - 2x$

2. $x - 2y = 0$
   $x + y = 1$

3. $y = -3x + 1$
   $y = \frac{1}{2}x - 6$

4. $y = 2x$
   $x = y + 4$

5. $2x + y = 3$
   $y = 4$

6. $9x - 7y = 6$
   $2x + 3y = 2$

II. Solve each system by the substitution method.

7. $2y = 5x$
   $y = x$

8. $-x - y = 5$
   $3x + 4y = 2$

9. $x - 4y = 8$
   $2y - 1 = x$

10. $x + 3 = y$
    $x - 3 = y$

11. $y = -3x$
    $y = 3x$

12. $2x - 3y = 4$
    $4x + 5y = 6$

III. Solve each system by the addition method. Indicate whether each
     system is independent, inconsistent, or dependent.

13. $x + y = 3$
    $2x + 2y = 6$

14. $x + y = 3$
    $3x + 3y = 6$

15. $x = y$
    $y = 3x + 1$

16. $x - y = 1$
    $x - y = -1$

17. $5x - 2y = \frac{1}{4}$
    $20x - 8y = 1$

18. $3x - 4y = 5$
    $2x + y = 3$

19. $x = y - 1$
    $2y = 3x - 1$

20. $4x - 3y = 5$
    $2x + 5y = 6$

♦♦♦♦♦

CHAPTER IX - TEST B

I. Graph each inequality.

1. $y \geq \frac{1}{4}x$
2. $x < 3y$
3. $y > -x + 5$
4. $2x - 3y < -5$
5. $x > -5$
6. $y < -2$

II. Graph each system of inequalities.

7. $x \leq -3$
   $y < -2$

8. $x > -5$
   $y \leq x + 1$

9. $y \leq 2$
   $x + y \geq 3$

10. $y \geq 3x$
    $3y \leq x$

11. $5y + 1 > 0$
    $2y - 3 \leq 0$

12. $x > -y$
    $2x > -3y$

13. $x + y \geq 2$
    $2x - 3y \leq -1$

14. $5x > y$
    $y \leq 5x$

15. $x < -2y$
    $2y + x \geq 1$

III. Solve each of the following.

16. The sum of the two numbers is 30 and their difference is 6. Find the numbers.

17. Five frames and two drawings cost $16.00; two frames and three drawings cost $13.00. How much does each cost?

18. The difference of two numbers is 17. The sum of half one of the numbers and four times the other is -14. Find the numbers.

19. Ten apples and eight lemons cost $2.80. Fifteen apples and three lemons cost $3.30. Find how much each costs.

20. The combined age of Joe and Pete is 42. Two times Joe's age is 15 more than Pete's age. How old are they?

◆◆◆◆◆

CHAPTER IX - TEST C

I. Complete the ordered pairs for each equation.

1. $3x + 2y = 8$ $(0, \quad)$, $(\quad , -\frac{1}{2})$, $(1/3, \quad)$
2. $x - y = 3$ $(-3, \quad)$, $(\quad , 0)$, $(\quad , -4)$
3. $4x - 5y = 2$ $(\quad , 2)$, $(3, \quad)$, $(-1, \quad)$
4. $x = y + 3$ $(\quad , -1)$, $(-3, \quad)$, $(0, \quad)$
5. $\frac{1}{4}x - \frac{1}{2}y = \frac{1}{4}$ $(0, \quad)$, $(\quad , -1)$, $(\quad , -3)$

II. Solve the following system.

6. $3x - y = 5$
   $x + 5y = 1$

7. $x - y = 1$
   $2x - y = 3$

8. $2x - 3y = 4$
   $2x = 1$

9. 2x = 3y
    3x = 2y

10. 9x + y = 0
    x + y = 1

11. 3x - 4y = -1
    x + y = 1

12. 2x - y + 3 = 0
    y - x - 1 = 0

13. 2y - 2x = 1
    x = y

14. (1/3)x + (2/3)y = 1
    2x - y = 1

III. Graph the inequalities.

15. x > -2
    y ≤ 3x - 1

16. y ≤ -1
    x > y + 1

17. y > 2x + 3
    x ≤ 2y - 4

18. x > 3
    y ≤ -2

19. y ≤ -x + 1
    x > 2y - 1

20. x - y < 0
    2x - y ≥ 2

♦♦♦♦♦

CHAPTER IX - TEST A - SOLUTIONS

I. Solve by graphing.

1. $y = \frac{1}{2}x + 3$ -- (1)
   $y = 1 - 2x$ -- (2)

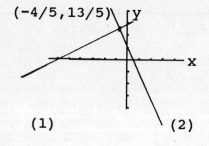

2. x - 2y = 0 -- (1)
   x + y = 1 -- (2)

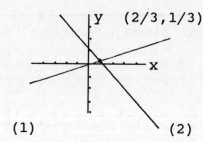

3. y = -3x + 1 -- (1)
   $y = \frac{1}{2}x - 6$ -- (2)

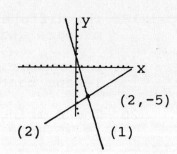

4. y = 2x -- (1)
   x = y + 4 -- (2)

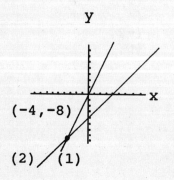

5. 2x + y = 3 -- (1)
   y = 4 -- (2)

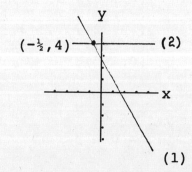

6. 9x - 7y = 6 -- (1)
   2x + 3y = 2 -- (2)

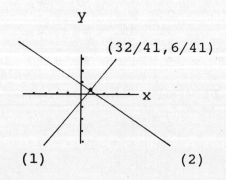

II. Solve each system by the substitution method.

7. 2y = 5x   -- (1)
    y = x   -- (2)

    y = x   -- from eq.(2)
    2[x] = 5x -- substitute y = [x] into eq.(1)
    3x = 0   -- simplify
    x = 0, y = 0
    (0, 0) is the solution to the system.

8. -x - y = 5   -- (1)
    3x + 4y = 2   -- (2)

    x = -5 - y   -- from eq.(1)
    3[-5-y] + 4y = 2 -- substitute [x = -5-y] into eq.(2)
    -15 -3y + 4y = 2 -- simplify
    y = 17, x = -22
    (-22, 17) is the solution to the system.

9. x - 4y = 8   -- (1)
    2y - 1 = x   -- (2)

    x = 4y + 8   -- from eq.(1)
    2y - 1 = [4y + 8]   -- substitute x = [4y + 8] into eq.(2)
    2y = -9   -- simplify
    y = -9/2, x = -10
    (-10, -9/2) is the solution to the system.

10. x + 3 = y -- (1)
    x - 3 = y -- (2)

    $m_1 = m_2 = 1$
    There is no solution to the system.
    (The two lines are parallel).

11. y = -3x -- (1)
    y = 3x -- (2)

    y = -3x   -- from eq.(1)
    [-3x] = 3x -- substitute y = [3x] into eq.(2)
    6x = 0   -- simplify
    x = 0, y = 0
    (0, 0) is the solution to the system.

12. 2x - 3y = 4 -- (1)
    4x + 5y = 6 -- (2)

    x = 2 + (3/2)y   -- from eq.(1)
    4[2 + (3/2)y] + 5y = 6 -- substitute x = [2 + 3/2y] into eq.(2)
    11y = -2   -- simplify
    y = -2/11, x = 19/11
    (-2/11, 19/11) is the solution to the system.

III. Solve each system by the addition method. Indicate whether each system is independent, inconsistent, or dependent.

13.   x + y = 3 implies y = -x+3   14. x + y = 3 implies y = -x+3
         2x + 2y = 6 implies y = x+3   3x + 3y = 6 implies y = -x+2
         This system is dependent.   This system is inconsistent.

15.   x = y -- (1)
         y = 3x + 1 -- (2)

         -y = -x   -- multiply eq.(1) by -1
    +) y = 3x + 1   -- from eq. (2)
         ——————————————
            0 = 2x + 1

          x = -½,   y = -½
        (-½, -½) is the solution.

16.   x - y = 1 implies y = x-1
         x - y = -1 implies y = x+1
         This system is inconsistent.

17.   5x - 2y = ¼   implies y = (5/2)x - 1/8
         20x - 8y = 1 implies y = (5/2)x - 1/8
         This system is dependent.

18.   3x - 4y = 5   -- (1)
         2x + y = 3   -- (2)

         8x + 4y = 12   -- multiply eq.(2) by 4
    +) 3x - 4y = 5   -- from eq. (2)
         ——————————————
         11x = 17

         x = 17/11, y = -1/11
        (17/11, -1/11) is the solution.

19.   x = y - 1   -- (1)
         2y = 3x - 1   -- (2)

         -2x + 2y = 2   -- multiply eq.(1) by 2
    +) 3x - 2y = 1   -- rearrange eq.(2)
         ——————————————
         x = 3

         y = 4   -- from eq.(1)
        (3, 4) is the solution.

20.   4x - 3y = 5 -- (1)
         2x + 5y = 6 -- (2)

         -4x - 10y = -12   -- multiply eq.(2) by -2
      +) 4x -   3y = 5   -- from eq.(1)
         ──────────────────
              -13y = -7

         y = 7/13
         x = (43/26)
         (43/26, 7/13) is the solution.

                        ◆◆◆◆◆

CHAPTER IX - TEST B - SOLUTIONS

I. Graph each inequality.

   1. y ≥ ¼x   2. x < 3y   3. y > -x + 5

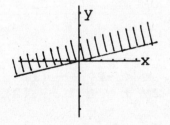   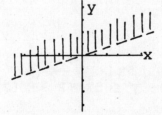   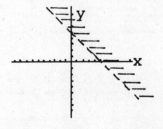

   4. 2x - 3y < -5   5. x > -5   6. y < -2

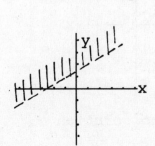   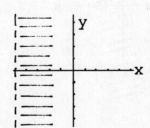   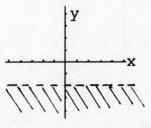

II. Graph each system of inequalities.

7. x ≤ -3 -- (1)
   y < -2 -- (2)

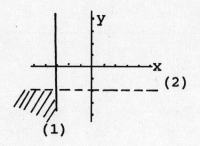

8. x > -5 --(1)
   y ≤ x + 1 -- (2)

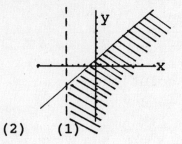

9. y ≤ 2 -- (1)
   x + y ≥ 3 -- (2)

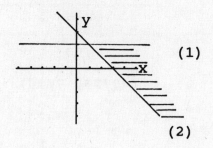

10. y ≥ 3x -- (1)
    3y ≤ x -- (2)

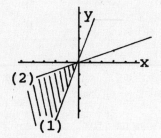

11. 5y + 1 > 0 -- (1)
    2y - 3 ≤ 0 -- (2)

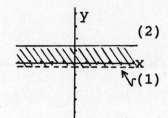

12. x > -y --(1)
    2x > -3y -- (2)

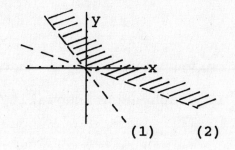

13. x + y ≥ 2 -- (1)
    2x - 3y ≤ -1 -- (2)

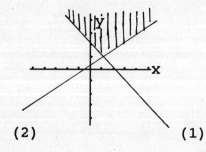

14. 5x > y -- (1)
    y ≤ 5x +2 -- (2)

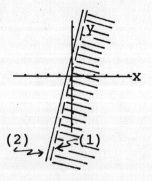

15. x < -2y -- (1)
    2y + x ≥ 1 -- (2)

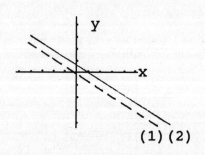

no solution

244

III. Solve each of the following.

16. The sum of two numbers is 30 and their difference is 6. Find the numbers.

    Let x represent one number and y represent the other.
    x + y = 30 -- the sum of two numbers
    x - y = 6   -- their difference
    2x = 36   -- add the two equations
    x = 18, y = x - 6 = 18 - 6 = 12.
    The two numbers are 12 and 18.

17. Five frames and two drawings cost $16.00, two frames and three drawings cost $13.00. How much does each cost?

    Let x represent the cost of a frame and y the cost of a drawing.
    5x + 2y = 16 -- (1)
    2x + 3y = 13 -- (2)

       15x + 6y = 48 -- multiply eq.(1) by 3
    +) -4x - 6y = -26 -- multiply eq.(2) by -2
       _____

       11x = 22

       x = 2, y = 3
    One frame costs $2.00 and one drawing costs $3.00.

18. The difference of two numbers is 17. The sum of half one of the numbers and four times the other is -14. Find the numbers.

    Let x represent one number and y represent the other.
    x - y = 17   -- (1)   the difference of two numbers
    ½x + 4y = -14 -- (2)
    x = 17 + y   -- from eq.(1)
    x + 8y = -28   -- simplify eq.(2)
    [17 + y] + 8y = -28 -- substitute x = [17 - y] into eq.(2)
    9y = - 45
    y = -5, x = 12
    The two numbers are 12 and -5.

19. Ten apples and eight lemons cost $2.80. Fifteen apples and three lemons cost $3.30. How much does each cost?

    Let x represent the cost of an apple and y the cost of a lemon.
    10x + 8y = 2.8 -- (1)
    15x + 3y = 3.3 -- (2) divide both sides by 3
    5x + y = 1.1
    y = 1.1 - 5x --- simplify the above equation
    10x + 8[1.1 - 5x]= 2.8 -- substitute y = [1.1 - 5x] into eq.(1)
    - 30x = - 6.0
    x = 0.2, y = 0.1
    The cost of an apple is $0.2 and a lemon is 0.1.

20. The combined age of Joe and Pete is 42. Two times Joe's age is 15 more than Pete's age. How old is each of them?

Let x represent Pete's age and y represent Joe's age.
x + y = 42   -- (1)
2y - 15 = x -- (2)
x = 42 - y   -- from eq.(1)
2y - 15 = [42 - y] -- substitute x = 2y - 15 into eq.(2)
3y = 57   -- simplify
y = 19, x = 2y - 15 = 2(19) - 15 = 23 -- from eq.(2)
Pete is 23 years old and Joe is 19 years old.

◆◆◆◆◆

HAPTER IX - TEST C - SOLUTIONS

I. Complete the ordered pairs for each equation.

1.	3x + 2y = 8	(0, 4),	(3, -½),	(1/3,7/2)
2.	x - y = 3	(-3, -6),	(3, 0),	(-1, -4)
3.	4x - 5y = 2	(3, 2),	(3, 2),	(-1,-6/5)
4.	x = y + 3	(2, -1),	(-3,-6),	(0, -3)
5.	¼x - ½y = ¼	(0, -½),	(-1,-1),	(-5, -3)

II. Solve the following system.

6.   3x - y = 5 -- (1)
       x + 5y = 1 -- (2)

       x = 1 - 5y -- from eq.(2)
       3[1 - 5y] - y = 5 -- substitute x = [1 - 5y] into eq.(1)
       -16y = 2
       y = -1/8, x = 1 - 5y = 13/8
       (13/8, -1/8) is the solution to the system.

7.   x - y = 1 -- (1)
       2x - y = 3 -- (2)

       x = 1 + y -- from eq.(1)
       2[1 + y] - y = 3 -- substitute x = [1 + y] into eq.(2)
       y = 1, x = 2
       (2, 1) is the solution to the system.

8.   2x - 3y = 4 -- (1)
       2x = 1   -- (2)

       x = ½ -- from eq.(2)
       2[½] - 3y = 4
       3y = -3
       y = -1
       (½, -1) is the solution to the system.

9.   2x = 3y -- (1)
        3x = 2y -- (2)

        -6x + 9y = 0 -- multiply eq.(1) by -3
    +)   6x - 4y = 0 -- multiply eq.(2) by 2
        _____
               5y = 0

        y = 0, x = 0
        (0, 0) is the solution to the system.

10.   9x + y = 0 -- (1)
        x + y = 1   -- (2)

        -x - y = -1 -- multiply eq.(2) by -1
    +) 9x + y = 0 -- from eq.(1)
        _____
        8x = -1

        x = -1/8, y = 9/8
        (-1/8, 9/8) is the solution to the system.

11.   3x - 4y = -1 -- (1)
        x + y = 1   -- (2)

        -3x - 3y = -3 -- multiply eq.(2) by -3
    +) 3x - 4y = -1 -- from eq. (1)
        _____
             -7y = -4

        y = 4/7, x = 3/7
        (3/7, 4/7) is the solution to the system.

12.   2x - y + 3 = 0 -- (1)
        y - x - 1 = 0   -- (2)

        2x - y + 3 = 0 -- from eq.(1)
    +) -x + y - 1 = 0 -- rearrange eq.(2)
        _____
        x + 2 = 0

        x = -2, y = -1
        (-2, -1) is the solution to the system.

13.   2y - 2x = 1 implies y = x + ½
        x = y   implies y = x
        m1 = m2 = 1
        Two equations are parallel (Inconsistent)

14. $(1/3)x + (2/3)y = 1$ -- (1)
        $2x - y = 1$ -- (2)

        $x + 2y = 3$ -- multiply eq.(1) by 3
    +) $4x - 2y = 2$ -- multiply eq.(2) by 2
        ─────────────
        $5x \quad = 5$

        $x = 1, \quad y = 1$
        $(1, 1)$ is the solution to the system.

III. Graph the inequalities.

15. $x > -2$ -- (1)
     $y \leq 3x - 1$ -- (2)

16. $y \leq -1$ -- (1)
     $x > y + 1$ -- (2)

17. $y > 2x + 3$ -- (1)
     $x \leq 2y - 4$ -- (2)

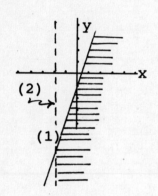
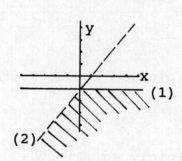
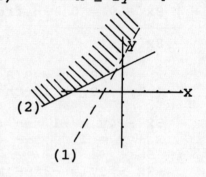

18. $x > 3$ -- (1)
     $y \leq -2$ -- (2)

19. $y \leq -x + 1$ -- (1)
     $x > 2y - 1$ -- (2)

20. $x - y < 0$ -- (1)
     $2x - y \geq 2$ -- (2)

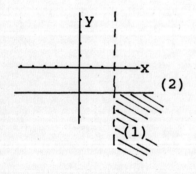
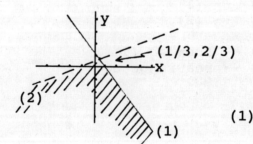
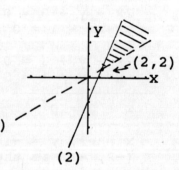

♦♦♦♦♦